Cambridge Studies in Historical Geography 7

HUMAN TERRITORIALITY

Cambridge Studies in Historical Geography

Series editors:
ALAN R. H. BAKER J. B. HARLEY DAVID WARD

Cambridge Studies in Historical Geography encourages exploration of the philosophies, methodologies and techniques of historical geography and publishes the results of new research within all branches of the subject. It endeavours to secure the marriage of traditional scholarship with innovative approaches to problems and to sources, aiming in this way to provide a focus for the discipline and to contribute towards its development. The series is an international forum for publication in historical geography which also promotes contact with workers in cognate disciplines.

1 Period and place: research methods in historical geography. *Edited by* ALAN R. H. BAKER *and* MARK BILLINGE
2 The historical geography of Scotland since 1707: geographical aspects of modernisation DAVID TURNOCK
3 Historical understanding in geography: an idealist approach LEONARD T. GUELKE
4 English industrial cities of the nineteenth century: a social geography R. J. DENNIS
5 Explorations in historical geography: interpretative essays. *Edited by* ALAN R. H. BAKER *and* D. J. GREGORY
6 The tithe surveys of England and Wales ROGER J. P. KAIN *and* HUGH C. PRINCE
7 Human territoriality: its theory and history ROBERT DAVID SACK

HUMAN TERRITORIALITY

Its theory and history

ROBERT DAVID SACK
Professor of Geography, University of Wisconsin

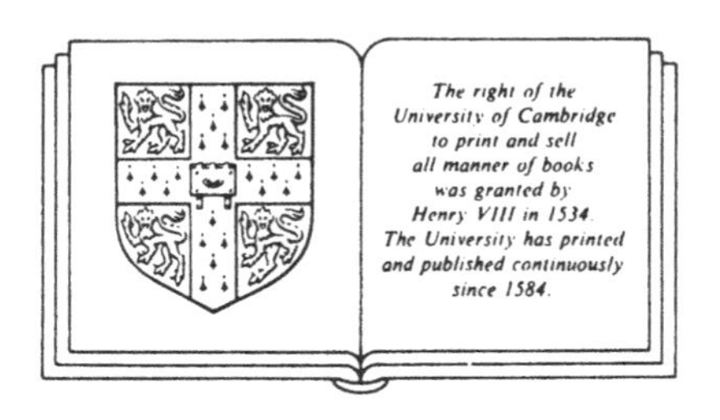

CAMBRIDGE UNIVERSITY PRESS
CAMBRIDGE
LONDON NEW YORK NEW ROCHELLE
MELBOURNE SYDNEY

CAMBRIDGE UNIVERSITY PRESS
Cambridge, New York, Melbourne, Madrid, Cape Town, Singapore, São Paulo, Delhi

Cambridge University Press
The Edinburgh Building, Cambridge CB2 8RU, UK

Published in the United States of America by Cambridge University Press, New York

www.cambridge.org
Information on this title: www.cambridge.org/9780521311809

First published 1986
Re-issued in this digitally printed version 2009

A catalogue record for this publication is available from the British Library

Library of Congress Cataloguing in Publication data

Sack, Robert David.
Human territoriality: its theory and history
(Cambridge studies in historical geography; 7)
1. Geography, Historical. 2. Human territoriality.
I. Title. II. Series
G141.S22 1986 911 85–30812

ISBN 978-0-521-26614-7 hardback
ISBN 978-0-521-31180-9 paperback

Contents

Illustrations

Acknowledgments

I wish to thank Alan Baker, Martin Cadwallader, Nick Entrikin, Brian Harley, Robert Ostergren, Karen Sack, Jane Schulenburg, Harold Schultz, Yi-Fu Tuan, and David Ward for reading the manuscript or portions of it. Their comments and encouragement were invaluable. I am indebted to Julie Fisher and Michael Solot for their help as project assistants, and to Linda Fuss for her skill and patience in preparing the manuscript. I also wish to thank the Fulbright program for funding a Fulbright Senior Research Fellowship to Leicester, England (1979–80) where the idea for this book began, the University of Wisconsin Graduate School for providing summer and research support for 1984, and I would like to acknowledge that portions of this material are based upon work supported by the National Science Foundation under Grant SES–8117807.

Introduction

Human territoriality is a powerful and pervasive element in our lives, but serious scholarship has only skirted its perimeters. This book attempts to help redirect research towards the core of territoriality by conceiving of it as an often indispensable means to power at all levels: from the personal to the international. Since the subject is vast and the uses of territoriality so varied, I can do no more in a single volume than offer a sketch and hope that the picture will serve to stimulate further research. In order to prepare the reader for the approach taken in this book, I would like to say a few words about the contexts in which I think a fruitful discussion of territoriality does and does not belong.

Perhaps the most well-publicized statements on human territoriality have come from biologists and social critics who conceive of it as an offshoot of animal behavior. These writers argue that territoriality in humans is part of an aggressive instinct that is shared with other territorial animals. The view presented in this book is quite different. Although I see territoriality as a basis of power, I do not see it as part of an instinct, nor do I see power as essentially aggressive. The power that a parent exercises over a child may be for the child's good, and that power may or may not be territorial. A parent may decide it is safer to keep the child indoors and away from the wet and cold of the rain. Keeping the child at home, as we shall see, is a territorial restraint. It may be a convenient strategy, but it is not the only means of keeping the child warm and dry. The parent could allow the child to play outside if he is well bundled up in warm rain gear.

Humans can use territoriality for a variety of often abstract reasons, few if any of which are motivations for animals. In fact, because territoriality in humans supposes a control over an area or space that must be conceived of and communicated, one can argue that territoriality in this sense is quite unlikely in most if not all animals. Territoriality in humans is best understood as a spatial strategy to affect, influence, or control resources and people, by controlling area; and, as a strategy, territoriality can be turned on and off. In geographical terms it is a form of spatial behavior. The issue then is to find out under what conditions and why terrioriality is or is not employed.

This book will examine human territoriality in the context of human motivation. But there remains the fact that the popular image of territoriality is drawn from works emphasizing biological links. Thus the efforts at drawing attention away from such a connection can be confounded by using a term that connotes them. Despite such drawbacks to the term territoriality (and that it is not a pretty sounding word), I have not been able to find a better one. Sovereignty, property, and jurisdiction are too restricted in scope to be suitable alternatives. Although an awkward term, I will use territoriality and trust that its former connotations will not draw attention away from what I believe to be its true signification: *a human strategy to affect, influence, and control.*

Territoriality in humans is best thought of not as biologically motivated, but rather as socially and geographically rooted. Its use depends on who is influencing and controlling whom and on the geographical contexts of place, space, and time. Territoriality is intimately related to how people use the land, how they organize themselves in space, and how they give meaning to place. Clearly these relationships change, and the best means of studying them is to reveal their changing character over time. Territoriality thus lies squarely within two geographical traditions: social geography and historical geography. We can of course claim that these are really interconnected and form a single approach – a social–historical one. Whereas few would deny that the two should be linked, and some have been able to incorporate both traditions in their work, it is not easy to combine both to everyone's satisfaction. The problem centers on the complex differences between the particularistic approaches of historical geography and the generalizing approaches of social geography and its theoretical component – spatial analysis. The differences between the particular and the general are confronted over again in the rest of social science.

Historical geography, which is closely allied with history, tends to undertake detailed examinations of places at certain periods. It may employ generalizations from social geography and other social sciences and it may arrive at general descriptions of people and society, but its primary focus is an understanding of the particular relationships that pertained at a particular place during a particular period. It is often 'long' on facts and descriptions, but 'short' on theory. In terms of the philosophy of geography, it tends to be 'ideographic.' Social geography (and its most generalizing component – spatial analysis), on the other hand, is closely allied to such systematic social sciences as economics, sociology, and political science, and tends to form abstract models of social–geographical relations and to test them usually in contemporary settings, though occasionally data from the past are used. In geographic terminology these approaches are called 'nomothetic.'

Clearly these can be conceived of as forming a continuum, and thus they need not in principle represent very different approaches. History and historical geography can use socal theories and help reformulate them, and social theories can be made more precise and pertinent under the scrutiny of historians. Yet in practice – due perhaps to personal preferences in research, to styles of analysis, to gaps between fact and theory – the continuum has been a bit thin in the middle. Historians and historical geographers often criticize systematic social science models as a-historical and claim that when the models are tested on the past, we learn very little about the period because these over-generalized models, rather than the historical contexts, select the facts to be explained. Social geographers and social scientists counter that many historical geographers and historians are too unwilling to generalize and to accept the fact that even detailed descriptions must be based on generalizations about behavior and about the past. And of course, when one tries to bridge these differences by practicing in the middle of the continuum, one runs the risk of not satisfying either end.

So it is within the tradition of human geography, and somewhere between the traditions of social and historical analysis, that this work on territoriality lies. The following contains both theory and history with perhaps a heavier emphasis on the former because my training has been in the spatial-analytic part of social geography. By theory I do not mean the full-blown positivistic conception of a series of nomothetic relationships linked together axiomatically and which can be used to predict human actions. Rather by theory I mean an interrelated group of characteristics which can be used to explain or make sense of behavior. This more flexible meaning is intended to suggest less than the positivistic ideal, but more than just some loosely connected notions. An important characteristic of territorial theory is that it is designed to disclose potential reasons for using territoriality. Which ones are used in fact depend on the actual context. Some of the reasons or effects will be used in practically any situation, and others will be used only under particular contexts. In this respect, the theory is phrased generally or abstractly drawing on social structure, but its specification and exemplification depends on particular historical context and on individual agency. The purpose of the book is not simply to test, exemplify, or illustrate generalizations. It is hoped that the book will also deepen our understanding of certain historical contexts by demonstrating how and why territoriality is used. Territoriality can shed light especially on the rise of civilization and on critical facets of modernity.

Territoriality then is an historically sensitive use of space, especially since it is socially constructed and depends on who is controlling whom and why. It is the key geographical component in understanding how society and space are interconnected. In exploring these issues, the book not only uses the past to illustrate the theory but also reconstructs parts of the history of territoriality in order to shed more light on past and present social organizations. But in combining theory and history, the book makes no pretense at disclosing new historical facts or sources. Rather it attempts to place old and well-known facts in a different light.

1

The meaning of territoriality

Territoriality for humans is a powerful geographic strategy to control people and things by controlling area. Political territories and private ownership of land may be its most familiar forms but territoriality occurs to varying degrees in numerous social contexts. It is used in everyday relationships and in complex organizations. Territoriality is a primary geographical expression of social power. It is the means by which space and society are interrelated. Territoriality's changing functions help us to understand the historical relationships between society, space, and time.

This book explores some of the more important changes that have occurred in the relationships between society and territoriality, from the beginning of history to the present. It does so by analyzing the possible advantages and disadvantages that territoriality can provide, and considering why some and not others arise only at historical periods. Exploring the advantages and disadvantages leads us to the theory of territoriality. Exploring when and why these come to the fore constitutes the history of territoriality and its changing relationships to space and society.

The history of territoriality and territoriality's relationship to space and society are informed by the theoretically possible advantages that territoriality can be expected to provide. After introducing the meaning of territoriality in this chapter we will explore in Chapter 2 the theoretically possible advantages of territoriality. The subsequent chapters will consider how and when these advantages are used historically and the effects they have on social organization. Chapter 3 will sketch the major changes in the relationships between territory and society from primitive times to the present and focus on the most important periods: the rise of civilization and the rise of capitalism. Chapter 4 will analyze the pre-modern development of territoriality within a complex organization – the Catholic Church. Chapters 5 and 6 will consider the development of territoriality in the modern period: Chapter 5 will explore the rise of the four-hundred-year-old political territorial organization of North America; Chapter 6 will explore the development of territoriality within work environments for the same span of time.

These periods and contexts are selected to illustrate the most important historical developments in the uses of territoriality. They will permit us to see that some territorial effects are universal, occurring in practically any historical context and social organization, that others are specific to particular historical periods and organizations, and that only modern society tends to use the entire range of possible effects. Exploring how modern society employs this range and especially why it employs territorial effects that were not of use to pre-modern societies, will help to unravel the meanings and implications of modernity and the future role of territoriality.

Examples of territoriality

Before we consider territoriality's theory and history, we must first describe what it is and what it does. To familiarize ourselves with the range of our subject, let us sketch territorial uses in three contexts. The first concerns the Chippewa Indians of North America and their contact with Europeans and serves to illustrate differences in territorial uses between pre-modern and modern societies. The second concerns territoriality in the modern home and the third considers territoriality in the modern work place. Both explore contemporary territorial uses in familiar small-scale contexts and point to the ubiquity of territoriality in modern life.

The Chippewa

Consider the group of American Indians, called the Chippewa (Ojibwe), who, in the early days of European contact, occupied a large area surrounding the western half of Lake Superior.[1] The Chippewa belong to the Algonkuian language group which covered much of the north central and north eastern sections of the United States and the south central and eastern portions of Canada. There were well over 20,000 Chippewa at the time of first European contact. Although the Chippewa possessed a common language, culture, and system of beliefs, they did not possess a central political organization. They were more of a collection of bands than a 'tribe.'

The Chippewa were primarily hunters, gatherers, and collectors. They lived on berries, nuts, roots, wild rice, fish, and game. Those who lived in the south and west portions of Lake Superior in areas having approximately 100 frost-free days or more per year were able to supplement their diets by cultivating corn and squash. Their material artifacts included canoes, bows and arrows, spears, traps, and baskets; and their shelters ranged from wooden tepee-like constructions to leantos and dugouts. Some within the community were better able than others to make these artifacts, but knowledge of how to construct them was available to all. Those who had superior abilities were looked upon as leaders. Leadership was earned. A leader would not impose his decision on his people and could not prevent a person from obtaining a livelihood. In economic terms, these people were egalitarian.

The size of Chippewa social units beyond the family varied seasonally. During the spring, summer, and early autumn, when berries, roots, wild rice, and fish were readily at hand and the larger game were plentiful, families would gather together to form a village of perhaps 100 to 150 people. During the winter months, when food was scarce, the families would normally disperse into smaller units, with an individual household occasionally going it alone. Even though single families could survive a season by themselves, they were rarely out of reach of others during the winter, and in the warmer months reconstituted their villages to undertake those numerous cultural and economic activities that required sustained cooperation. When together, band members hunted, gathered, and shared their produce. Friendships were established and marriages planned. Membership in bands seems to have been voluntary. If tensions arose, or if needs changed, a family could leave one band and join another.

What can be said about Chippewa territorial organization? It is clear that as an entity the Chippewa occupied a vast area. But their habitation was never clearly bounded and fluctuated from year to year. On the east the Chippewa were interspersed among the closely related and friendly Ottawa and Potawatomi; in the north they were intermingled among the normally friendly Cree; in the west with more Cree, and Assiniboin.[2] The Chippewa had their greatest difficulty with the eastern and prairie Dakota who were along their southern and western frontiers. But a large tract of unoccupied no-man's land provided a buffer zone between them and their Dakota neighbors. Even if the perimeter of the Chippewa 'nation' had been stable, it is doubtful that it would have been circumambulated by a single Chippewa, or that many among them would possess a map-like representation of their collective domains.[3]

Chippewa bands, too, occupied particular areas, but their sites shifted after several years as did their social compositions. A band's encampment at a particular site and its use of the resources of the surrounding area must have been known and accepted by neighboring bands. But this does not mean that a band needed to claim a specific territory exclusively for its own use and defend it against incursions by other Chippewas. Population was sparse enough and food abundant enough so that when a band used an area it is unlikely it would be to the exclusion of confreres. Individuals and families within these egalitarian bands did not themselves 'own' land. The land was the community's to use, and band members were allowed to share in its use. A band could apportion part of an area to a particular family, but this did not mean the family owned the land or excluded others from it. This applies to the use of land for agriculture as well as for hunting and gathering.

The growing season north of the Great Lakes was too short for the Chippewas there to practice agriculture, but south and west of Lake Superior the cultivation of corn and squash formed an important supplement to Chippewa diet. These Indians had their fields nearby their villages. Each family may have had its own garden which it cleared, planted, and tended alone, or the process may have been collective. In any case these gardens were not clearly demarcated and fenced-in territories.

At the time of European contact, then, these people were hardly territorial as a 'nation,' although they may have been occasionally territorial as individual bands or as families within bands. Yet even here their assertion of control over an area was often imprecise, seasonal, and strategic. Bands or families may have laid claim to an area only if they were reasonably confident that the resources they were after would be there and if they knew there would be competition for these resources from other groups. Imagining these very conditions to predominate allows us to consider how a group 'such as' the Chippewas might alter and intensify their territorial use. We say 'such as' because some of the factors we will consider, although important causes of changes in territorial use in other pre-literate societies, and although present in Chippewa society, were not in fact the primary ones to alter Chippewa territorial use. Yet entertaining them as possibilities will help us understand how in general a simple pre-literate society can develop primarily internal pressures to alter relationships between territoriality and social organization.

In this vein suppose that game becomes scarcer and for those Chippewa in the south more time must be devoted to agriculture. Suppose also that for some in Minnesota and Wisconsin the horse becomes part of their culture. Members of the community may still collectively clear the fields, plant, and tend crops, but how are these now vital crops to be protected from the wild animals, from the very young children, and from the horses? It is possible that these are minor difficulties and that no special precautions are needed. The threat from wild animals may be negligible; the adults can closely supervise the children and their access to the crops; and the horses may find enough grass to graze on so that they will not forage in the gardens. But it could also be the case that even if these are not serious problems, the community finds it more convenient either to fence off the fields or to fence in the horses, or both. The purposes of these clear territorial demarcations would be to establish different degrees of access to things in space. Yet little else need change. The community may still maintain its original goals.

But it is not difficult to have our imaginations go a step further to consider conditions of greater crowding, making unavoidable more complex territorial partitions within the band. The size of the community itself may grow to the point where casual community work efforts become unmanageable, and population pressure from other groups may make it impossible for a family simply to leave one band for another. Even though the community may still be egalitarian – even though the land is still the community's – the fields may now be allocated to families on the basis of need, and family plots may have to be demarcated and access restricted simply to prevent inadvertent trampling. The possibilities for territoriality can multiply within this egalitarian society. But there is a point at which some of these possibilities may actually interfere with the values of community sharing and cooperation. This is not to say that different uses of territoriality alone can transform social relations from, in this case, an egalitarian to a class structured society. But territoriality can be a catalyst in the process of change and can be used differently and to as much advantage by a class divided as by an egalitarian society. If for example a Chippewa ruling family were to emerge claiming access to some or all of the community's resources, territoriality would be an extremely useful device to affect its claims.

These speculations point to the possibility of territorial changes occurring largely from forces within the society. Such transformations have in fact been documented for several pre-literate societies and will be examined more closely in a subsequent chapter. But for the Chippewa, most of the social and territorial transformations were imposed upon them by European and American economy and polity.

The European fur trade soon strained social relationships within the bands. It strained egalitarian and communal efforts. It affected hunting habits and an ecology of the area, and it may have increased individual and family territorial control at the expense of communal access. But the adoption of private property was selective. Some have claimed that as a result of the fur trade individual families among Woodland Indian tribes, including the Chippewa, appeared to own hunting grounds that were passed down from father to son. But upon close inspection of the evidence it seems that private territorial control may have been exercised only over access to furs and not to other resources. According to Leacock these hunting territories, at least for the Montagnais, 'did not involve true land ownership. One could not trap near another's line, but anyone could hunt game animals, could fish, or could gather wood, berries, or birchbark on another's grounds as long as these products of the land were for use, and not for sale.'[4]

European settlement east of the Alleghenys also increased population pressure throughout the upper Mid-West as tribes moved farther west to find new land. Population pressure and reliance on trade further strained communal social-territorial relationships of bands; many families became both dependent on and skillful in the fur trade. This adaptation actually helped extend the Chippewa domain until by the 1840s Chippewa settlement

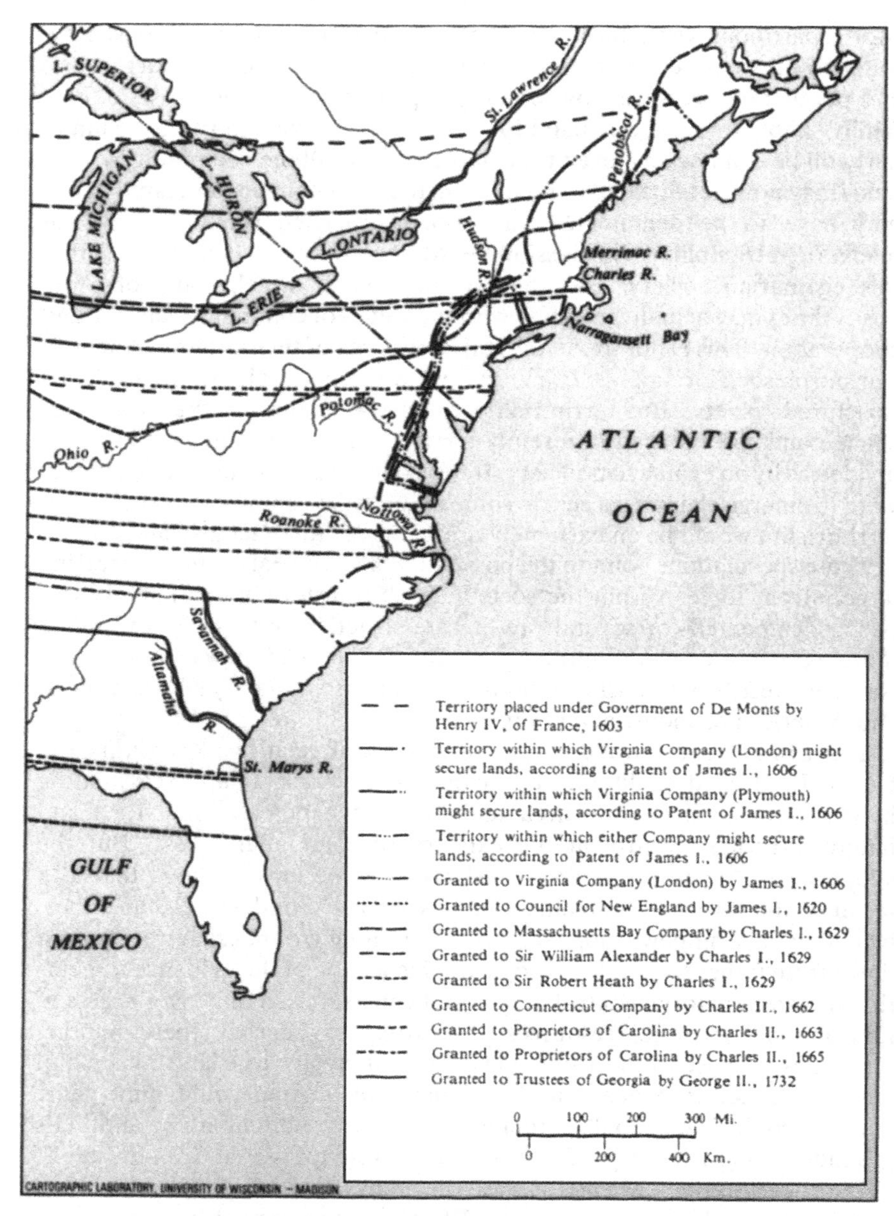

Figure 1.1 Colonial land grants, 1603–1732
Source: Atlas of the Historical Geography of the United States, ed. Charles O. Paullin and J. K. Wright (Westport, CT: Greenwood Press, 1932).

covered the areas of western Lake Superior, east to the shores of Lake Huron, north practically to the shores of Hudson Bay, west to Lake Winnipeg, and south to central Minnesota and Wisconsin.

But the most far reaching effect on Chippewa social and territorial organization came from the imposition by the Europeans of hierarchical and territorial political jurisdiction. Unbeknownst to the Chippewas and to the original colonizers, the early English grants and charters forming the territorial units of the colonies (as seen in Figure 1.1) made claim to much of the Chippewa area.

The Virginia Charter of 1609 itself encompassed the entire upper Mid-West. But these early claims were not enforceable. From the mid-seventeenth century to the mid-eighteenth, the upper Mid-West was nominally under French control until it was signed over to the English and their colonies in 1763. Shortly thereafter, Virginia, Massachusetts, Connecticut, and New York made claims to parts of this land. High among the priorities of the First Congress of the United States was the disposition and government of the North West Territory: the approximately 170 million acres west of the Ohio to the Mississippi. By 1786, and after much wrangling, the claimants had ceded all of these lands to the United States, and a series of ordinances, based on Thomas Jefferson's 1783–4 proposal, and culminating in the ordinances of 1787 and 1796, provided procedures for the governance of this territory. The plan, illustrated by areas 1–5 in Figure 1.2, was to divide the North West into not less than three and no more than five states and to admit each into the Union once it had a population of 60,000. In addition the land would be surveyed according to a regular rectangular grid, the units of which would help delineate state boundaries, form the entire boundaries for counties and townships, and boundaries of saleable parcels of land[5] (see Figure 1.3).

These were plans, written and mapped out on paper, for a land virtually unknown to Europeans, and which, at its farthest reach, was over 1,000 miles from the Eastern Seaboard where the decisions were being made. With the stroke of a pen, Americans of European descent were to classify, divide, and control people, including Chippewas, solely on the basis of their location in space. This imposition of territory had both a social and an economic dimension. At the social level a national, state, or local boundary could apportion a society among a number of jurisdictions. The political units to which the Chippewas belonged changed frequently as the territorial map of the North West took shape. Eventually part of the Chippewa were to be in Canada, another part in Minnesota, another in Michigan, and yet another in Wisconsin.

With the exception of reservations, most of the Chippewa land within the United States was sold as private parcels lying within particular territories or states, and after the states were admitted into the Union, the political

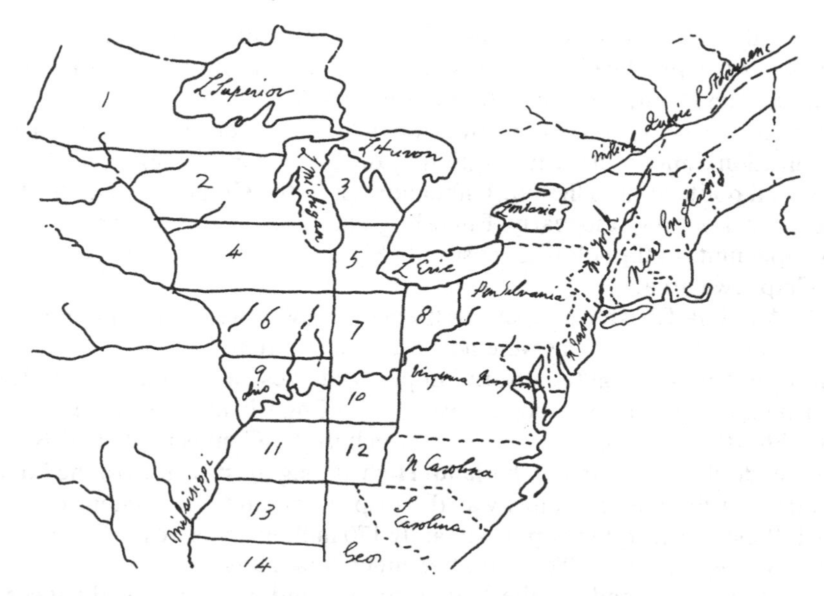

Figure 1.2 Jefferson–Hartley map (1783) of proposed states
Reproduced by permission of the William L. Clements Library, University of Michigan.

territorial partitioning continued along county and township lines, further subdividing and segmenting former Indian lands. These local units formed political communities for non-reservation Indians as well as for European settlers. The reservations were 'permanent' territorial molds containing land that Europeans found least desirable. As though retaliating against this territorially imposed restriction on Indian culture, the boundaries of reservations often formed an impediment to the neat geometric symmetry of neighboring townships, cutting through the rectangular land survey and interrupting the domains of local authority.

These newly imposed political territories (national, state, county, and township) were designed to serve the needs of the white man's market-oriented society. While the imposed boundaries segmented older communities, they forged newer and different ones geared to a dynamic market system. And within this system territorial partitioning became a primary vehicle for defining property. Unlike the aboriginal Indian's communal use of land, the white man used territory to partition land into saleable parcels. Each piece of private property was a territory under the control of an individual. Each had a monetary value and each could be bought and sold again and again.

The different functions that white men and Indians gave to territoriality

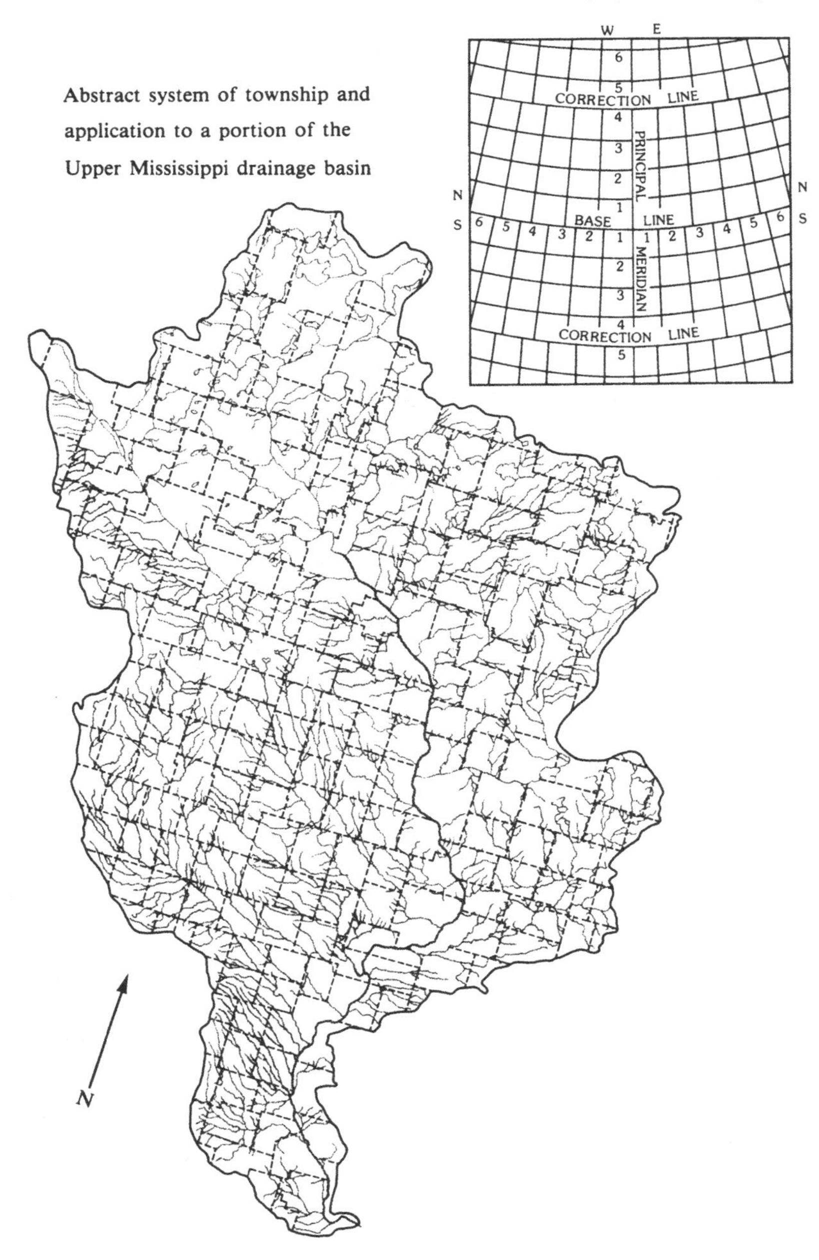

Figure 1.3 United States rectangular land survey
Source: Hildegard B. Johnson, *Order upon the Land* (Oxford, 1976). Reproduced by permission of the Delegates of Oxford University Press.

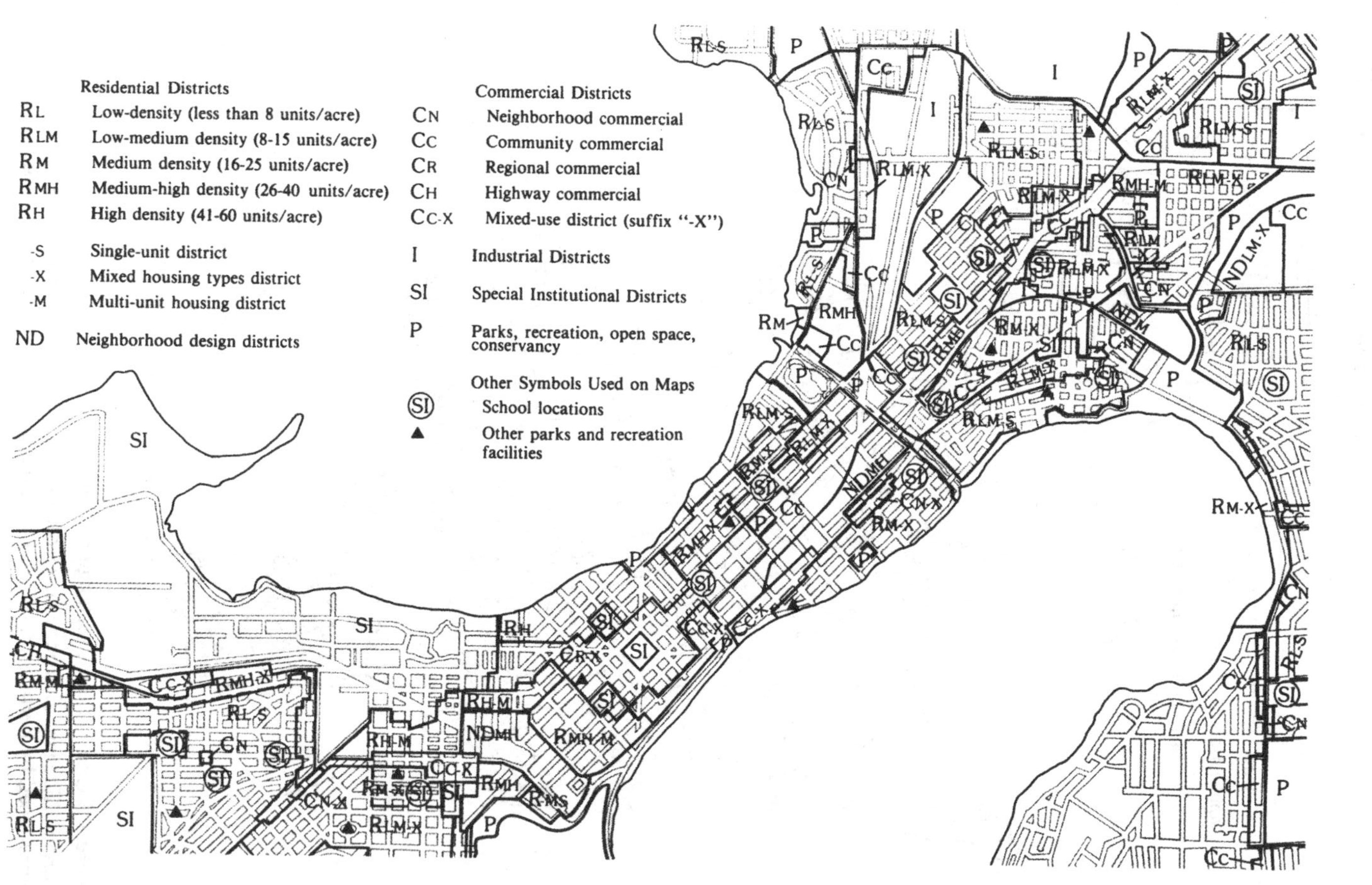

Figure 1.4 Land use map for portion of Madison, WI (1978)

sheds light on their long troubled relationship. But it also serves to illustrate how the establishment and uses of territory are intermeshed with social-historical contexts. The accurate lines on political maps and land surveys of the white man were possible to construct because his society was literate and capable of printing, surveying, and determining longitude and latitude. But more fundamental is the fact that these territories were created and used to support his complex hierarchical society which was based on private property and which used territory to define and organize its own membership. In contrast to a Chippewa, who was born into a Chippewa community and was accepted socially and culturally by the Chippewa people, a Wisconsinite was simply someone who resided within the boundaries of Wisconsin. In modern Western culture, simply living within a territory often enables one to be a member of a community.

The areas once occupied by migrating Chippewa bands are now virtual mazes of nested and overlapping hierarchies of political, quasi-political, and private territories (see Figure 1.4). Although not always visible on the ground they have precise and fixed boundaries specified in maps and documents and affect numerous segments of our lives. Simply by being located at one moment in place 'x,' say an upper Mid-Western city, one is either on a piece of public or private property. In either case, one is automatically grouped along with others in the same location as being situated within the jurisdiction of police district 'a,' fire district 'b,' sanitation district 'c,' school district 'd,' planning district 'e,' state court district 'f,' federal court district 'g,' the city of 'h,' the county of 'i,' and the state of 'z.' Shift your position just the slightest bit and you will change your relationship to one or more of these units as well as your relationship to others.

Establishing political jurisdiction and delimiting private ownership of land are the most familiar uses of territoriality in the Western World. However, territoriality has and continues to play important roles in other aspects of social relations. Let us briefly consider two modern examples: territoriality within the home and within the work place. Each can be used to specify further territoriality's meaning and illustrate its interconnections to social contexts.

The home

Consider a twentieth-century North American parent who is a property owner on former Chippewa land. He is at home, doing housework and minding his two very young children. As the parent is dusting and vacuuming it dawns on him that the children are in the kitchen 'helping' to wash the dishes. The difficulty is that the well-intentioned young helpers are perilously close to dropping the plates. Their activities are taking place in space. In geographic terms, they are 'spatial.' Although the surroundings are different, the parent is presented with a problem that is much like the one the Chippewa parents would have faced if they had been concerned that their children would trample the fields. Geographically speaking the North American (and the Chippewa) parent has only two strategies to prevent disaster. He can have a face to face, heart to heart, talk with the children, thanking them for their efforts but explaining that there may be difficulties if they continue. He might also remove the dishes from their reach. (The Chippewa parent could not have removed the plants.) In either undertaking the parent is attempting to control the spatial actions of his children, and what they have access to in space, by focussing on the specific objects of the actions, like the dishes (or the plants).

The intent is to alter the children's access to things in space but, in the above, territoriality is not being invoked. Territoriality, as the second strategy, is brought to bear when the parent decides simply to restrict the children's access to things in space by telling them 'they may not go into the kitchen without permission' (or that they may not enter the fields unsupervised). That is, the kitchen (or the field) is now 'off limits.' Here the parent is attempting to limit the children's access to things by asserting *control over an area.*

Notice that the kitchen (or the field) is there all the time. It is a bounded place. In the non-territorial case it simply was not demarcated as an area of control. In the second case it was. In other words, a place can be a territory at one time and not at another, and a territory can create a place where one did not before exist. Moreover, the assertion of territoriality may apply only for a limited time. The modern parent may have said 'don't go into the kitchen *now* while I am vacuuming.' Or the territorial restrictions can be lifted when the objects the parent wishes to protect are now in the cupboard 'out of reach.'

The kitchen is imbedded in other places that are also territories; the house, the city, the state. The authority of these territories was not directly invoked in this case, but was in the background and could be drawn upon in other situations that could arise even in the kitchen. Note also that declaring the kitchen to be off limits to the children and enforcing the assertion is not the end of the matter. The parents' assertions have to be stated clearly to the children, they must be able to understand them, and their behavior has still to be monitored. All of these tasks involve further behavior in space. Using territoriality may help reduce some types of spatial interactions, the amount of monitoring, and the destruction of plates. But if the assertions of territorial control cannot fail, then the alternative is a non-territorial 'spatial' strategy. If the children persist in entering the kitchen and touching the plates, the parent may physically have to remove the children. In geographic terms, territoriality is a form of spatial interaction that influences other spatial interactions, and requires non-territorial actions to back it up.

Within the context of child rearing and the home, the aboriginal Chippewa and the North American parent would possess similar choices about the role of territoriality. But the choices are quite different in the context of work. Most people in North America now work in a place that is under someone else's control.

The work place

Suppose that the same North American parent is employed as a secretary in a modern office building. Typically these places contain large rooms filled with desks and typewriters. Each desk is designed as a 'work station.' The secretary is employed to type, and part of the working agreement is that he be in the office at his desk for a specified number of hours per day, five days a week, fifty weeks a year. The modern secretary may leave the work station. But if he does so often and without permission he may be in violation of the work agreement and stand to lose his job. Even if the secretary is permitted to leave his work station his movements within the buildings are likely to be restricted. He cannot simply wander into any office. Perhaps the only areas to which he is free to go are those designed for 'traffic' such as hallways and corridors, and those open to 'workers' such as lavatories and coffee rooms. For the secretary, territoriality acts as a physical restraint.

After work hours, say five p.m., the territorial functions of the building become 'inverted.' The secretary leaves for home and instead of restraining and molding his actions, the building is now off limits to him and to the public. Present at day but dissolved at night are the internal territorial partitions of offices and work stations that separated workers and levels of personnel. The building may still be occupied, but this time by janitors and night watchmen who, unlike all but the highest levels of management, have access to practically every part of the building.

More eventful changes can occur in the territoriality of the office building. The firm using the building may move or go bankrupt and the building itself may be demolished. More far reaching in its geographical effects is the possibility that with modern tele-communications systems the office as a territory may become obsolete because much if not all of the secretary's work can be done in any place, even in the home. This may make it unnecessary for people to gather in one place called the office. Yet this does not eliminate work territories entirely. There still needs to be restricted access to the place where work and equipment will be located, even if that is in the house, and employers may very well have to enter the house to check on the worker and the equipment he is using. What has changed is the form of territorial organization and its relationships to non-territorial spatial relations.

Notes on meanings

It is clear from these examples that territoriality covers a wide range of activities for which there are often other perhaps more richly descriptive names. Calling rooms, buildings, property rights in land, political sovereignty, and legal jurisdictions over area, as well as roads and cities, 'territories,' serves no purpose unless the term enhances our understanding of these particulars. Viewing familiar activities as territorial should add to our understanding of them. This means that territoriality must be defined broadly enough to cover these and other cases and yet richly enough to illuminate its different effects. We need to know not only what territoriality is, but what it does. It is principally on helping to point to the important effects of a phenomenon that the value of a definition rests.

A definition is never all inclusive. It focusses on one or a few of a phenomenon's characteristics. A phenomenon that contains these characteristics fits the definition, yet it will also possess many other characteristics, and can go by other names as well. A statue can be a work of art, an investment, a record of a human likeness, a piece of marble, and a mass. Each in its turn contributes to our understanding of the statue's uses and effects. An apple, too, is many things, most of which are different from the statue. But the two have things in common. They both take up space, and they both have weight or mass. Knowing what their masses are means that mass is clearly enough defined so that it can be observed in even very different types of phenomena.

But a concept or term needs to be more than clear. It needs to point to connections with other attributes. In this sense mass is not only a clear concept but a useful one.[6] Knowing the mass of an object can tell us much about the phenomenon's actual and potential connections with its environment. We would know for instance how strong a floor must be to support the statue and the apple. We would be able to anticipate the impact each would have if dropped from a two story window. Knowing the mass of something broadens our understanding of it and its connections to the world. But by no means does it tell us all there is to know about the object and its interrelationships. The statue and the apple are both masses. But they exhibit many other clearly definable and significant attributes which they do not share and which cannot be disclosed by viewing them as instances of mass.

The same applies to territoriality. It is one thing to define territoriality clearly so that a room, a home, a field, an office, and a city are seen to be instances of territoriality. It is another to have our understanding of these phenomena and their interrelationships deepened by examining them as territories. The latter condition occurs only if our meaning of territoriality is clear and rich enough to suggest how it is joined with other facets of behavior.

Defining territoriality

Territoriality, as simply 'the control of area,' has served so far as a shortened definition. But this description is neither precise nor rich enough to take us much farther. From our examples of the parent and the children, the secretary in the work place, and the members of the hunting–gathering society, it can be seen that territoriality involves the attempt by an individual or group to influence or affect the actions of others including non-humans. It is this important yet general effect that must be emphasized and which is elaborated in the following formal definition of territoriality. In this book territoriality will be defined as *the attempt by an individual or group to affect, influence, or control people, phenomena, and relationships, by delimiting and asserting control over a geographic area.* This area will be called the *territory.* Before we explore the significance of this definition, some further clarification of its domain is in order.[7]

Once again, it should be emphasized that a place can be used as a territory at one time and not at another; that is, in creating a territory we are also creating a kind of place. But it is important to distinguish between a territory as a place and other types of places. Unlike many ordinary places, territories require constant effort to establish and maintain.[8] They are the results of strategies to affect, influence, and control people, phenomena, and relationships. Circumscribing things in space, or on a map, as when a geographer delimits an area to illustrate where corn is grown, or where industry is concentrated, identifies places, areas, or regions in the ordinary sense, but does not by itself create a territory. This delimitation becomes a territory only when its boundaries are used to affect behavior by controlling access. For instance a formerly ordinary geographical place or region such as a corn belt or a manufacturing area may become designated by the government as a region to receive special financial assistance or as an area to be administered by a special branch of government. In this case the boundaries of the region are affecting access to resources and power. They are molding behavior and thus the place becomes a territory. By the same token, what geographers call nodal regions, market areas, or central place hinterlands are not necessarily territories. They can be simply descriptions of the geographic extent of activities in space. They become territories though if the boundaries are used by some authority to mold, influence, or control activities. Thus a chain of supermarkets may use market areas – the actual geographic limit of the drawing power of a supermarket – to define each supermarket manager's jurisdiction (i.e. his responsibilities for advertising).[9] A person or group can of course control more than one territory, and in modern society, many ordinary kinds of places must become territorial to exist as places.

Territoriality need not be defended area, if by that is meant that the area itself is the object of defense, and that the defender(s) must be within the territory defended. Territory can be used to contain or restrain as well as to exclude, and the individuals who are exercising control need not be inside the territory. Indeed, they need not be anywhere near it. A fence or wall can control, so too can a 'no trespass' sign. The definition points out that territoriality establishes control over area as a means of controlling access to things and relationships.

Territoriality is a strategy to establish different degrees of access to people, things, and relationships. Its alternative is always non-territorial action, and non-territorial action is required in any case to back it up. For example if the Chippewa decide to fence in their gardens, these fences must be maintained by direct physical labor; and if they should break, the children and the horses must be watched and guarded by direct non-territorial control. (If the children in the modern home continue to enter the kitchen even if the parent told them *not to*, then the parent will have to resort to a non-territorial form of intervention.)

Both the boundaries of a territory and the means by which they are communicated are not unalterable. Land holdings change size. So too do nation states. A boundary fence may be replaced by one of a different type such as a ditch. A child may recognize a threshold to a room as a boundary, or else have the door to the room closed. Most territories tend to be fixed in geographical space, but some can move. For instance the personal space or social distance surrounding a person travels with that person when he maintains the distance. The convention among ships of war not to come too near foreign naval vessels in the high seas is an example of a movable territory.

Territories can occur in degrees. A cell in a maximum-security prison is more territorial than a cell in a country jail, which is more territorial than a room in a half-way house. A closed classroom, with its desks anchored to the floor and its children seated all day at their desks, is more territorial than an open classroom which has no fixed seats for each child and that allows the children to move about from one activity to another. Degrees of territoriality are far more difficult to compare when selecting examples from different institutions and societies. Are activities of an automobile worker on an assembly line more territorially circumscribed than those of an office secretary in a secretarial pool? The finer points of measuring the intensity of territoriality will be addressed later on. For now it should be borne in mind that although we can make rough estimates of territorial intensity, difficulties arise when comparing one context with another.

Territoriality can be asserted in a number of ways. These include job descriptions (how long you must be seated, where you are and are not allowed to go, etc.), legal rights in land, brute force or power, cultural norms and prohibitions about the use of areas, and subtler forms of communication such as body posture. But once again, if the assertion is not clear and understandable then it is unclear whether territoriality is being exercised.

Definitions should be clear enough to point out that something does or does not fit the definition. But even a clear definition has fuzzy edges in practice. If I am in a library and place my books on an empty table, am I simply relieving myself of a burden or claiming a part of the table as mine? And if the latter, am I asserting control over an object, the table, or over a territory that the object circumscribes? There is no harm in admitting that borderline cases occur which can go either way. A definition can have some exceptions or fuzzy parts and still be useful especially when there are innumerable clear-cut examples that fall within its domain.

Considering territoriality as a strategy for differential access side steps the fruitless issue of whether human territoriality is in any sense biologically rooted.[10] By making it a strategy it places territoriality entirely within the context of human motivations and goals. Our definition of territoriality indeed cuts across perspectives and levels of analysis. It involves the perspectives of those controlled and those doing the controlling, whether they be individuals or groups. It draws upon physical, social, and psychological effects. This cross-cutting of other fields is not new to geography and is paralleled by the range of interconnections that have been developed in the rest of the field.[11]

The significance of territoriality

The formal definition of territoriality not only tells us what territoriality is but it suggests what it can do. This suggestion comes from three interdependent relationships which are contained in its definition. These three disclose the logic and significant effects of territoriality. First, by definition, territoriality must involve a form of classification by area. When someone says that anything, or even some things, in this room are his, or are off limits to you, or that you may not touch anything outside this room, he is using area to classify or assign things to a category such as his, or not yours. He need not define or enumerate the kinds of things that are his or are not yours. When using territoriality, the parent did not have to tell the children what they should not touch. They were simply not allowed in the room. According to Piaget, there are only two major forms of classification.[12] One is by type and the other is by area. Territoriality indeed can employ both but it always employs the latter.

Second, by definition, territoriality must contain a form of communication. This may involve a marker or sign such as is commonly found in a boundary. Or a person may create a boundary through a gesture such as pointing. A territorial boundary may be the only symbolic form that combines a statement about direction in space and a statement about possession or exclusion.

Third, each instance of territoriality must involve an attempt at enforcing control over access to the area and to things within it, or to things outside of it by restraining those within. More generally, each instance must involve an attempt at influencing interactions: transgressions of territoriality will be punished and this can involve other non-territorial and territorial action.

The logic of territoriality rests on the fact that the advantages of using it must be linked with one or more of these interconnected relationships. Because they are essential facets of territoriality the three must also be the basis for the significance of territoriality. It is simple to illustrate how each can be a reason for using territoriality. Consider the first characteristic: that territoriality involves a form of definition or classification by area. Definition by area can be extremely useful when either we cannot enumerate things, people or relationships we want to have access to, or when we wish not to divulge such a list. A football team practicing new plays before the big game may not want the opponents to know about them. To help keep them secret, the coach may use territoriality to exclude observers from the field and stands.

Consider the second characteristic: communicating by using a boundary. The boundary may be a simpler device for communicating possession than enumeration by kind. If the children in the kitchen are very young, they may have difficulty understanding which objects in the kitchen they are or are not allowed to touch. Territoriality may be the only means of conveying the parent's wishes to the children. This is especially the case if elsewhere and under different circumstances the children also are allowed to touch similar types of objects like dishes. Instead of presenting the children with a complicated rule about when handling dishes is not permissible, it is simply more direct to tell them they may not cross beyond this line, or enter/leave this room.

Consider the third characteristic: enforcement of access, in the context of the hunting Chippewa community. To make sure that the children would not trample the fields, it may be easier to fence in the children than to follow them around. Similar circumstances occur in our society. It is easier to supervise convicts by placing them behind bars than by allowing them to roam about with guards following them. Controlling things territorially may save effort.

These three facets of territoriality can be found in all societies, but they, in turn, generate further potential effects that can be equally important but which occur only in particular historical contexts. How this happens, and what the effects are, is somewhat technical and will be discussed at length in Chapter 2. For purposes of illustration, we can point out that by classifying at least in part by area rather than by kind or type, territoriality can help relationships become impersonal and can help mold future activities within a hierarchy. We noted that the Chippewa did not need to use territoriality to define their members, but the white man did. The primary definition of membership within a North American state or city is domicile within the political territory. This definition allows complete strangers to become members of the same community. Moreover, unlike the Chippewa community, the territory of the city acts as a container and as a spatial mold for other events. A city's influence and authority, although spreading far and wide, is legally assigned to its political boundaries. The territorial city becomes the object to which other attributes are assigned, as in the case of the political territory of the city being the unit receiving federal aid.

Fostering impersonal relationships and geographically molding activities within a hierarchy are but two of many identifiable consequences of the three facets of territoriality which follow from its definition. These and the others will be framed broadly enough in Chapter 2 to encompass the range of territorial uses and yet precisely enough to deepen our understanding of particular cases. The effects are potential because each one need not be employed in every instance of territoriality, and some have been used only at particular times in history. A suitably broad yet clear definition which points to the general implications of territoriality for humans is what has been missing in previous work on territoriality.

Previous approaches

Most of the considerable literature on territoriality is about animal behavior and does not concern us unless social scientists have borrowed from it in discussing human territoriality.[13] Though not as voluminous as the animal territorial literature, discussions of human territoriality are extremely varied and difficult to summarize. There are as yet no comprehensive reviews and what follows is not an attempt to provide one but rather is a brief illustration of some of the key problems from which many human territorial studies suffer.

Overall, previous analyses of human territoriality have been deficient in three important respects, with particular studies containing one or more of these deficiencies. First, in many cases researchers do not clearly distinguish the term territoriality from the term spatial. For them, simply having events take place in and through space is sufficient for them to have the action fall under the category territorial. Because these studies do not define territoriality as a particular kind of behavior in space, they miss the opportunity of offering a systematic analysis of territoriality. Any insights they present are difficult to attribute to territoriality in particular, and difficult to generalize about.[14] (There are even those which use the term figuratively to refer to 'cognitive' territories.)[15]

Closely related to the first are those studies which actually focus on examples of human territoriality without calling them such. Studies of zoning, of private property rights in land, of political sovereignty often do not recognize that their subject matter belongs to a territorial class of actions. Hence these studies miss important territorial implications.[16] In the case of the parent controlling his children, it is because we know at least three possible effects of territoriality that we are able to suggest that the parent resorted to territoriality because it made it unnecessary for him to identify, by kind, the things he wished to control. And it is this knowledge of territorial effects that make us expect this very use of territoriality to occur in other quite different contexts. Do not nation states use this effect when they declare sovereignty over everything and anything within their geographic domain? Parents and nation states do not, and probably cannot, list what it is they wish to control, and not listing what it is that is under control allows territoriality to hide what is being controlled. Consider simply the number of times parents hide things from children by not letting them enter places, or that states hide things from foreigners and even citizens by restricting entry to areas or regions within the country. Recognizing that territoriality is a general strategy for establishing access to things and pointing to its generally expected effects can help deepen our understanding of its use in particular cases.

Third are studies which have been conscious of isolating real territorial behavior in humans but which have the drawback of being far too narrow in their meaning. They may have focussed entirely on one social-geographical scale. This is the case in the social-psychological literature portraying territoriality as a form of personal space.[17] Other studies may be too narrow in the territorial effects they stipulate. For example some psychological studies view the use of territoriality by an individual as an expression of specific personality characteristics such as a desire for dominance or security.[18] Linking territoriality to particular needs occurs especially in studies which suppose that humans and animals use territoriality for the same essential biological reasons, i.e., as a means of obtaining food, mates, and controlling population size. Focussing on these narrow effects may make territoriality in humans appear to be something like an instinct rather than a strategy that can be turned on and off.[19] Moreover, these narrow meanings and their emphases on particular scales, purposes, or functions are often embodied in the formal definitions of territoriality. The social psychologist exploring only the personal level and psychological effects of territoriality may define it with this emphasis in mind. For instance, 'self' and 'personality' are part of Altman's definition of territoriality as 'a self/other boundary regulation mechanism that involves personalization of or making of a place or object and communication that it is "owned" by a person or group. Personalization and ownership are designed to regulate social interaction and to help satisfy various social and physical motives.'[20] There occurs the opposite problem of defining territoriality too generally. When it is intended to mean simply the control of area, we are left without any suggestion about purpose or intent, except that area itself is both an object and an end.[21]

Although these are among the major pitfalls in previous work on territoriality, a few researchers have avoided them, and many of those who have not, have none the less pointed to some of its important aspects. Rather than isolate the positive components piecemeal, they will be noted and incorporated in the work which follows.[22]

Territoriality and geography

Whereas work on territoriality has often been unwittingly about non-territorial spatial behavior, work in geography on spatial behavior has often ignored the territorial. In geography, both natural and human or cultural activities are called 'spatial' to remind everyone that they occur in space and have spatial properties such as locations, shapes, and orientations. Spatial analysis is the branch of geography interested in the interrelationships between activities in the landscape and their spatial properties. In human geography, these include not only the actual locations, extensions, and patterns of things, but how these are described and conceived of in different social and intellectual perspectives. (The identical landscape pattern may be described and evaluated economically, aesthetically, symbolically, and so on.) Geography's concern with multiple uses and conceptions of space, and with the historical geographies of different peoples, presents space as a complex framework in which individuals and groups are situated, through which they interact, and by which they make statements. Yet these interconnections between space and behavior rest on territoriality, the study of which has remained in the background, all but neglected by spatial analysis.[23]

The businesses, farms, and cities studied by geographers are not only places or locations in space with multiple meanings, but also occur and remain in place because there exist numerous social rules and regulations allowing some things to be in certain places and not others. Even the movements of peoples, goods, and ideas require society to set aside roads and the like for transportation and to disallow other activities from taking place on them. Modern city streets are designed for bicycles, cars, trucks, and buses and not for pedestrians. Highways are designed for traffic powered primarily by the internal combustion engine.

For the most part, people and their activities cannot find room in space without forms of control over area – without territoriality. The challenge is to show how and why this is the case. Unfortunately spatial analysts have not systematically explored territoriality to discover if there is a logic to territorial control in the same way as there has been an exploration into the question of whether there is a logic to non-territorial spatial organization and interaction. Instead, they have focussed on the objects territoriality has helped to form and support and have left territoriality – the geographical bonding agent – in the background.

Spatial analysts understand very well that activities compete for locations. In this respect the focus of their research has been on the process of selecting one site over another and the role played by distance or geographical accessibility in connecting sites. Emphasizing distance has led to a geographical logic based on the metrical properties of space. But spatial analysts have not seriously considered the possibility that geographical logic can be extended by the even more complex logic involved in territorial uses of space.[24] The logic of territorial action is more complex than the logic of distance because territoriality is imbedded in social relations. Territoriality is always socially constructed. It takes an act of will and involves multiple levels of reasons and meanings. And territoriality can have normative implications as well. Setting places aside and enforcing degrees of access means that individuals and groups have removed some activities and people from places and included others. That is, they have established different degrees of access to things.

Territoriality, then, forms the backcloth to human spatial relations and conceptions of space. Territoriality points to the fact that human spatial relationships are not neutral. People do not just interact in space and move through space like billiard balls. Rather, human interaction, movement, and contact are also matters of transmitting energy and information in order to affect, influence, and control the ideas and actions of others and their access to resources. Human spatial relations are the results of influence and power. Territoriality is the primary spatial form power takes.

Territoriality and history

Different societies use different forms of power. They have different geographical organizations and conceptions of space and place. Geographical landscapes and meanings change as societies change. Historical geography is concerned with these interconnections. Historical geography points to the socially historically dependent context of spatial organization and meaning; and territoriality points to the fact that geographical organization and meaning, while depending on many things, also presupposes the maintenance of different degrees of access to people, things, and relationships. Spatial organizations and meanings of space have histories and so too do the territorial uses of space; the three histories are indeed closely interrelated.

The logic of territoriality will show that, as a spatial strategy, it offers several advantages to help affect, influence, and control. These constitute the domain of reasons for, or consequences of, using territoriality. They explain how and why territoriality is being used and are the basis of its import. Whether or not particular advantages are used in a particular case depends on who is controlling whom and for what purposes. Some advantages can be expected to occur in practically any situation at any time. We found the modern parent employing territoriality in the kitchen because it made it unnecessary for him to explain what it was he did not want his children to handle. We can also imagine the Chippewa parent using territoriality for the same reason.

Not defining what it is that is under one's control is practically a universal advantage of territoriality. We can expect that other very important effects would appear in most types of societies, and still others would appear in only a few. For example, territoriality in the modern world is often an essential means of defining social relationships. As we pointed out, people who reside in a North American city are entitled to the public services of that city. Location within a territory defines membership in a group. This use of territory – to define belonging in a community – occurs to a somewhat lesser degree in pre-modern civilizations, but occurs hardly, if at all, in primitive societies where social relationships are so clearly and so strongly upheld. 'Primitives' may use territoriality to delimit and defend the land they occupy but they rarely use it to define themselves. Other effects of territoriality occur primarily in contemporary society, and still others would take place with equal importance in modern and pre-modern civilizations. As we shall see, the uses of territoriality have been cumulative. Primitive societies found need for a few. Pre-modern civilizations employed these and a few others, and modern society has employed virtually the entire range of possible effects.

We have mentioned a few of the possible territorial effects – the ease by which territoriality can classify, communicate, and enforce control, the ease by which it can define social relationships impersonally and hierarchically. What must be considered now is the range of theoretically possible effects and their interrelationships. These issues are addressed in the theory of territoriality.

2
Theory

We noted that the definition of territoriality contains three interrelated facets. Territoriality must provide a form of classification by area, a form of communication by boundary, and a form of enforcement or control. What will be argued now is that approximately seven other potential effects can be linked to these three facets, and that they, plus the original three, lead to approximately fourteen combinations of characteristics. Note that the precise number is not the critical issue. They can be collapsed into fewer than ten or fourteen. What is critical is that the definition of territoriality be rich enough to delimit the range of potential advantages offered by a territorial strategy and at a level of generality that is precise and useful. Specifying these effects, how they are connected to one another, and the conditions under which they would be employed constitute the theory of territoriality.

The theory will be presented in two parts. First, territoriality is conceptually abstracted from the multiplicity of social-historical contexts. This allows room to describe the internal logic of territoriality: to reveal the range of effects which constitute the domain of reasons for using territorial as opposed to non-territorial strategies and their logical interrelationships. Second, the theory hypothesizes that certain historical contexts will draw upon specific potential effects and, in a very general sense, matches historical contexts with territorial effects.

The theory is both empirical and logical. The first three tendencies are derived from the definition of territoriality. The others, while not entirely derivable from the definition, none the less are logically interconnected and linked to it. Calling the following analysis a theory does not mean that we are taking a mechanistic approach to people and their uses of territory. On the contrary, the theory will present the effects of territory as possibilities which range from the physical to the symbolic: a range spanned by the broad field of spatial analysis. Nor does the word theory mean that accurate predictions about territoriality can be made. Human behavior is far too 'open ended' to make possible precise social predictions of any consequence. Rather, by theory, is meant that we can disclose a set of propositions which are both empirically and logically interrelated and which can help make sense out of complex actions. In other words, the theory can help us understand and explain, but it is not likely to help us predict precisely what will happen in the future.

The complex structure of the theory may be more easily pictured if we resort to an analogy from physical science. Noting that the theory contains two parts – the range of effects and their uses in historical cases – we can say, at the risk of being branded mechanistic, that the first part is analogous to examining the 'atomic' structure of territoriality: the three facets (classification, communication, and enforcement) are its 'nucleus' and the ten primary and fourteen combinations of effects or tendencies are its 'valences.' These form the potential links that will be drawn upon if and when territoriality is used. The second part is analogous to placing territoriality in a periodic table of types of social-historical organizations and suggesting bondings that can be expected when these contexts use territoriality. Sketching the bonds between historical contexts and territorial effects will be the purpose of the subsequent chapters.

Before turning to the theory itself, a few more words about method and terminology are in order. Territoriality's effects are not simple relationships. Because they pertain to people, not to atoms, they are more appropriately termed potential 'reasons' or 'causes' of, or potential 'consequences' or 'effects' of, territoriality. The appropriateness of one set of names over the other depends on whether an individual (or group) is establishing new territories (in which case the appropriate couplet would be reasons/causes) or are using already existing ones (in which case the appropriate couplet would be consequences/effects). As to whether something is a reason or a cause, or a consequence or an effect, is impossible to know without looking closely at the specific case. And even then there are many who argue that little difference exists between the two. For the sake of simplicity, reasons, causes, consequences, effects, will be used interchangeably to show that these are applicable in any case; and the terms 'potentialities' or 'tendencies' will cover all four options.

Despite this effort at simplification, the theory is still complex and technical. We will need to describe ten *tendencies* and fourteen *combinations* of tendencies for a total of twenty-four effects. This is unavoidable. The theory must be developed as fully as possible early on since the historical chapters are organized around the claims of the theory. Each of the twenty-four effects will be given a common-sense name. In addition, to help distinguish the theory's internal structure, the first ten will each be assigned a number and the fourteen combinations a letter (e.g., the first tendency – classification – will be identified as *1*, and the third combination –

complex hierarchy – will be identified as *c*). The numbers and letters will be used only in this chapter for cross referencing. The remainder of the book will refer to the tendencies and combinations by their names. Omitting the letters and numbers may make it more difficult for the reader to be alerted by the fact that a particular tendency is being addressed. But using only the name of the effect will still allow the reader to identify it as such while making the theoretical structure less obtrusive and distracting to the narrative.

The social construction of territoriality

Conceptually isolating and describing territoriality to some degree apart from particular social contexts may seem analogous to the quest for the meaning of geographic distance in spatial analysis.[1] One critical difference is that territoriality is always socially or humanly constructed in a way that physical distance is not. It can be granted that, in a rudimentary sense, the act of conceiving, describing, and measuring distances is a matter of social construction, and so too are the social forces that place things in certain patterns in space. But territoriality is even more intimately involved with social context. Territoriality does not exist unless there is an attempt by individuals or groups to affect the interactions of others. No such attempt, nor indeed no interaction at all, need exist between two objects in space for there to be a specifiable distance between them.[2] Distances can be compared and measured, but there is little that can be said abstractly about their potentials to affect behavior. Their influence depends on there being actual channels of communication such as roads, railways, and the like, which contain these distances. Indiscriminate substitution of the physical measure of distance for the physically and socially significant channels of communication or interactions runs the risk of treating distance non-relationally.[3]

Unlike distance, territorial relationships are necessarily constituted by social contexts (however general) in which some people or groups are claiming differential access to things and to others. Because of this, more can be said abstractly about the effects of territoriality than can be said about distance, and yet, because territoriality is a product of a social context, whatever is said about it, no matter how abstract, can have normative implications affixed to it and thus can lead back to a social context. It is important to make clear that these normative implications refer to judgments people make about the uses of territoriality. An effect of territoriality may be considered by some as good, or neutral, or bad. Most may agree that using territoriality to prevent children from having access to plates in the kitchen can be an effective and even a benign strategy. Yet to a few it may be deceitful because the parent does not have to disclose to the children which objects they are not allowed to touch. The normative implications people assign to actions, and in this case territorial actions, are important parts of their effects. A parent may realize territoriality is efficient, but may not use it because he believes it is deceitful. The theory then must have room for the ethical and normative judgments that can be assigned by others to the uses of territoriality. This helps link the theory to society. Yet the theory itself will not present procedures by which one can judge whether an action is, on its own merits, good or bad.

When presenting the tendencies, the social contexts will be pushed far into the background (though specific examples are used to clarify their meanings, they should not be interpreted as specifying the context of the tendency) and the general normative implications will not be addressed until the combinations are discussed. Indeed, some of the combinations differ from one another in the degree to which they draw upon what others may label as benign or malevolent connotations. These normative terms are still intended to be very abstract and general. But by way of an illustration we may consider that a benign context to some may mean that a relationship is non-exploitative. Such a context might be approached at an individual level when a parent uses territoriality to prevent a young child from running into traffic; and at a group level when the workers of a democratically organized and controlled factory elect some of the members to serve for terms as managers. A malevolent territorial relationship, on the other hand, might be thought to occur when differential access through territoriality benefits those exercising territoriality at the expense of those being controlled.[4]

Keeping the descriptions of the tendencies neutral and the normative meanings of the combinations general helps to separate the expression of the theory of territoriality from particular theories of power and society. This allows territoriality an intellectual 'space' of its own and prevents territoriality from becoming the captive of any particular ethical theory or theory of power. The second part of the theory draws upon the capacity of these tendencies to have normative implications and thus to point to particular types of social contexts that may employ them. In this way the theory can be reconnected to specific historical cases and to theories of power.

Theory: part 1

Ten tendencies of territoriality

By definition, territoriality, as an assertion of control, is a conscious act, yet the person(s) exercising territoriality need not be conscious of the ten potentials or tendencies for these effects to exist. These tendencies of territoriality come to the fore, given certain conditions. Moreover, they are not independent of one another. In fact, the first three listed below – classification, communication, and enforcement – can be considered logically (though not empirically) prior. They are the bases by which the other seven potentialities of territoriality are interrelated, and any or all of the ten can be possible reasons for its use. Even if the first three are not important as reasons in some instances, they must nevertheless still be present because they are part of the definition. In other words, territoriality must provide classification, communication, and enforcement, but it can be 'caused' by one or several or all of the ten. Let us proceed in order from number *1* to number *10* and again be reminded that the terms used to describe them could apply to benign, neutral, or malevolent social contexts. (Each tendency is numbered and the italicized word(s) will serve as the name(s) of the tendency in subsequent chapters.) As the second section will show, tendencies are logically interconnected in numerous ways. The following is more a list of definitions of the tendencies than an illustration of their interrelationships. Yet the order in which they are discussed suggests how some lead to others.

1. Territoriality involves a form of *classification* that is extremely efficient under certain circumstances. Territoriality classifies, at least in part, by area rather than by type. When we say that anything in this area or room is ours, or is off limits to you, we are classifying or assigning things to a category such as 'ours' or 'not yours' according to its location in space. We need not stipulate the kinds of things in place that are ours or not yours. Thus territoriality avoids, to varying degrees, the need for enumeration and classification by kind and may be the only means of asserting control if we cannot enumerate all of the significant factors and relationships to which we have access. This effect is especially useful in the political arena, where a part of the political is its concern with novel conditions and relationships.
2. Territoriality can be easy to *communicate* because it requires only one kind of marker or sign – the boundary. The territorial boundary may be the only symbolic form that combines direction in space and a statement about possession or exclusion. Road signs and other directional signs do not indicate possession. The simplicity of territoriality for communication may be an important reason why it is often used by animals.
3. Territoriality can be the most efficient strategy for *enforcing* control, if the distribution in space and time of the resources or things to be controlled fall well between ubiquity and unpredictability. For instance, models of animal foraging have shown that territoriality is more efficient for animals when food is sufficiently abundant and predictable in space and time whereas non-territorial actions are more suitable for the converse situation. The same has been shown to hold in selected cases of human hunting and gathering societies.[5]
4. Territoriality provides a means of *reifying* power. Power and influence

are not always as tangible as are streams and mountains, roads, and houses. Moreover, power and the like are often potentialities. Territoriality makes potentials explicit and real by making them 'visible.'

5. Territoriality can be used to *displace* attention from the relationship between controller and controlled to the territory, as when we say 'it is the law of the land' or 'you may not do this here.' Legal and conventional assignments of behavior to territories are so complex and yet so important and well understood in the well-socialized individual that one often takes such assignments for granted and thus territory appears as the agent doing the controlling.
6. By classifying at least in part by area rather than by kind or type, territoriality helps make relationships *impersonal*. The modern city, by and large, is an impersonal community. The primary criterion for belonging is domicile within the territory. The prison and work place exhibit this impersonality in the context of a hierarchy. A prison guard is responsible for a block of cells in which there are prisoners; the guard's domain as supervisor is defined territorially. The same is true of the relationship between the foreman and the workers on the assembly line, and so on.
7. The interrelationships among the territorial units and the activities they enclose may be so complicated that it is virtually impossible to uncover all of the reasons for controlling the activities territorially. When this happens territoriality appears as a general, *neutral*, essential means by which a place is made, or a space cleared and maintained, for things to exist. Societies make this *place-clearing function* explicit and permanent in the concept of property rights in land. The many controls over things distributed in space (as the interplay between preventing things without the territory having access to things within and vice versa) become condensed to the view that things need space to exist. In fact, they do need space in the sense that they are located and take up area, but the need is territorial only when there are certain kinds of competition for things (in space). It is not competition for space that occurs but rather a competition for things and relationships in space.[6]
8. Territoriality acts as a *container* or *mold* for the spatial properties of events. The influence and authority of a city, although spreading far and wide, is 'legally' assigned to its political boundaries. The territory becomes the object to which other attributes are assigned, as in the case of the political territory being the unit receiving federal support.
9. When the things to be contained are not present, the territory is conceptually 'empty.' Territoriality in fact helps create the idea of a socially *emptiable place*. Take the parcel of vacant land in the city. It is describable as an empty lot, though it is not physically empty for there may be grass and soil on it. It is emptiable because it is devoid of socially

or economically valuable artifacts or things that were intended to be controlled. In this respect, territoriality conceptually separates place from things and then recombines them as an assignment of things to places and places to things. As we shall see, this tendency can be combined with others to form an extremely important component of modernity – that of emptiable space.

10. Territoriality can help *engender more territoriality* and more relationships to mold. When there are more events than territories or when the events extend over greater areas than do the territories, new territories are generated for these events. Conversely, new events may need to be produced for new and empty territories. Territoriality tends to be space-filling.

These are brief descriptions of the ten consequences that we hypothesize could come from the use of territorial organization and that would be drawn upon to explain the reasons for having territorial, as opposed to non-territorial, activity. Once again, these tendencies are not independent and their precise number and definition is not as critical as the question of whether or not they circumscribe the domain of its potential effects. Not all of them need be used in any particular territorial instance in history, and (as mentioned) their meanings or imports would depend on the specific historical conditions of who controls whom, how, and for what purpose. Some of their interconnections were noted and more will become apparent as we discuss the primary combinations.

Primary combinations

Most of human behavior occurs with hierarchies of territorial organizations: individuals live in cities, which are in states, which are in nations. People work at desks which are in rooms, which are in buildings. Hence everything we said about territories applies, in addition, to hierarchical territorial organizations. For example, having territory used as a mold within a hierarchy of territories, as it is in the context of municipalities, states, and the nation, could mean that a goal, such as 4 percent unemployment, can be assigned precisely for one geographical level, such as the national, rather than for another, such as the state or the municipal. Assigning tasks or responsibilities to different territorial levels may become a general political strategy. Territoriality, as a means of circumscribing knowledge and responsibility, can be used to assign the lowest level and the smallest territory the least knowledge and responsibility and the highest level and largest territory the most.

In this vein, and still without being specific about social contexts, we can proceed to illustrate more of the logical interrelations among the tendencies

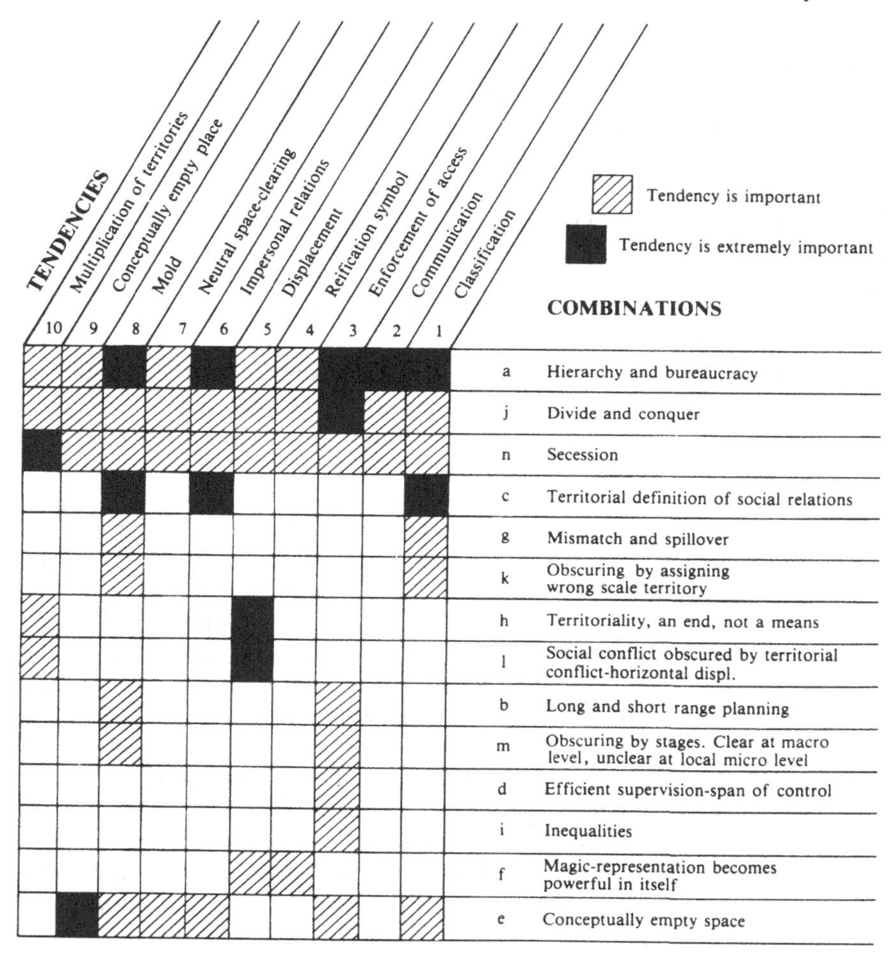

Figure 2.1 Internal links among tendencies and combinations

by considering the possible primary combinations and their general import within social hierarchies. We will begin with a list, as we did with the tendencies. This list will address the relationships of tendencies to combinations within hierarchies. The order of the list suggests how some of the combinations lead to others, but the major interconnections among the combinations will be discussed later.

Figure 2.1 is a matrix tracing the connections among the elementary tendencies (*1–10*)) to form the primary combinations (*a* to *n*). The array of combinations in Figure 2.1 is not alphabetical as is the order in the list of combinations. This is because Figure 2.1 groups the combinations discussed in the list by their mixtures of tendencies rather than by the interconnections among the combinations. (The dynamics among the combinations are illustrated in Figure 2.2.)

Figure 2.1 shows only the important links. A darkened square signifies that a potential is extremely important, and a striped one that it is moderately important. A blank means that the tendency is not important for that particular combination. It does not mean that it has no effect at all. (Note again that whereas (*1*), (*2*), and (*3*) must be attributes of territoriality, they need not be important causes/consequences of territoriality. Their inclusion in the matrix is to indicate when they, as characteristics of territoriality, also become important consequences of territory.) Without linking territoriality to specific social contexts, it is impossible to be more precise about the degree to which each tendency contributes to a combination or whether a darkened area can be called a necessary and/or sufficient condition. It should not be forgotten that some combinations differ only in the connotations and weights placed on their tendencies.

a. Perhaps the most general combination is that all ten tendencies can be important components of complex and rigid *hierarchies*. (*1*), (*2*), (*3*), (*6*), and (*8*) are especially important, for they can allow hierarchical circumscription of knowledge and responsibility, impersonal relationships, and strict channels of communications, all of which are essential components of *bureaucracy*. The strength of (*8*), impersonal relations, affects the degree to which the bureaucracy is modern, according to Weber's criterion.[7]

b. Not only can the scope of *knowledge* be graded according to territorial levels, but so too can the scope of *responsibility* in space and time through enforcing (*3*) and molding (*8*) access to information. *Long-term planning* could be made the responsibility of the highest level which would have access to the greatest knowledge and responsibility and *short-term planning* (or no planning at all) would be the responsibility of the lowest territorial level. Moreover, an action could be subdivided into stages: the first having to do with overall initiation of policy and the last having to do with carrying out of the details. The first would pertain to the higher territorial levels, the last to the lower levels.

c. Upper echelons of a hierarchy tend to use territories to define (*1*), enforce (*3*) and mold (*8*) groups, with the result that members may be collected and dealt with impersonally (*6*). It is this cluster (*1*, *3*, *6*, and *8*) to which the historical-anthropological literature points when it discusses the *territorial definition of social relationships*.[8] This is a relative concept and its opposite is a *social definition of territory*. The difference between them is a matter of degree. A relatively extreme case of a territorial definition of social relations can be found in our previous comparison of membership in a twentieth-century North American community as compared with membership in the Chippewa community. Yet even in North America both territorial and social definitions can be found in the same place. The requirement for receiving police, legal, and fire protection from an American municipality is that one be located within the geographical bounds of that community. Those who do not even reside there but simply pass through receive these benefits. On the other hand, within the same city, being a visitor in someone's house does not make the visitor a part of the household and does not give the visitor the right to use the household's resources. An actual claim to territoriality may involve elements of both, as when full political citizenship in American municipalities, although granted on the basis of residence, is still given only to U.S. citizens. As we noted in the Chippewa example, primitive societies rely almost entirely on a social definition of territoriality whereas civilizations and especially modern societies do the opposite. Continuous and intense territorial definitions lead, as we shall see, to a conceptually emptiable space (*3*).

d. A significant yet simple combination is that the hierarchical territorial circumscription (*3*) of knowledge and responsibility can provide a very efficient means of supervision. For example, constraining the movements of prisoners by placing them in cells makes easier the task of supervising them than if they were allowed to roam freely in the prison. Indeed, even a prison without cells but with an outside wall provides a more effective means of supervision than a non-territorial form of contact such as handcuffing a prisoner to a guard. An important quantitative index of the degree of supervisory efficiency would be the *span of control*, i.e., the number of supervisors per supervisees. This measure is a well-known index of organizational structure and is exhibited by all territorial organizations.[9]

e. The combination of elements constituting a territorial definition of social relationships (*1*, *3*, *6*, and *8*) in conjunction with a neutral space-clearing device (*7*), and especially a conceptually empty place (*9*) point to the possibility, on a practical level, of continually filling, emptying, and rearranging things in a territorial mold for the purpose of efficient functional control. This constant manipulation of things within a territory would lead, on an abstract level, to a conceptual separation and recombination of things and space and thus to a *conceptually emptiable space*. Space – not just place as in (*9*) – would appear as an efficient functional framework for events. Events and space would seem to be only contingently related. This possibility is especially significant in modern society and characterizes the conception of territory most closely linked with modern modes of thought. Science, technology, and capitalism make practical the idea of repeatedly and efficiently 'filling' and 'emptying' and moving things about within territories of all scales. Planners expect states to lose or gain population year by year, and federal support to states allows for such changes. On a smaller scale, factory buildings serve as territorial molds or containers to house first one industry, then another, or when no one rents the building, to contain nothing at all. Geographical mobility and territorial power at the political level, and emptying, filling, and arranging at the architectural level, loosen the bonds between events and location and present territory and space as a background for the occurrence of events, a background that can be described abstractly and metrically. Changes in activities are especially prevalent in modern culture. Consumer society makes change essential. Geographically, change and the future are seen as sets of spatial configurations different from those that exist now or that existed in the past. A place that has not changed its appearance has been bypassed by time; it has stood still. Planning for change and thinking of the future means imagining different things in space. It involves imagining the separation and recombination of things in space. Territoriality serves as a device to keep space emptiable and fillable.[10]

f. The combinations of reification (*4*) and displacement (*5*) could lead to a *magical* mystical perspective. Reification through territory is a means of making authority visible. Displacement through territory means having people take the visible territorial manifestations as the sources of power. The first makes the sources of power prominent, whereas the second disguises them. When the two are combined they can lead to a mystical view of place or territory. This often occurs within religious uses of space. For example, Catholicism reifies when it makes the distinction between the primary sources of power (i.e., faith and the Church invisible) and the physical manifestations of these (i.e., the Church visible). But Catholicism displaces when it has worshippers believe that the physical structures of the Church and its holy places emanate power. The same relationships occur in nationalism. The territory is a physical manifestation of the state's authority, and yet allegiance to territory or homeland makes territory appear as a source of authority.[11]

g. The territorial component in complex organizations can have a momentum of its own, on the one hand increasing the need for hierarchy and bureaucracy and on the other diminishing their effectiveness. This can come about when definition by area (*1*) leads (unintentionally) to the circumscription of the wrong area or the wrong scale and thus to a *mismatch* of territory or a *spillover* of process. The mismatch may become aggravated by using the territory as a mold (*8*). Mismatch and spillover would diminish the organization's effectiveness; but because knowledge and responsibility within the organization are unequally shared, responsibility for rectifying the problem may fall to the existing hierarchy and thus entrench and even increase the role of bureaucracy.

h. Displacement (*5*) and territorial multiplication (*10*) make it easier for the territory to *appear* to be the *end rather than the means* of control. (Appear is emphasized because territoriality, as a strategy, is always a means to an end.) The Catholic Church offers an example of this. By the fifth century A.D., the powers of archbishops were measured in part by the numbers of dioceses and parishes under their control. In order to increase this power, an archbishop would subdivide his see and thereby increase the number of bishops and priests under his supervision.

i. The territorial component can have a *momentum* of its own *to create inequalities*. Its facility in helping to enforce differential access to things (*3*) can become institutionalized in rank, privilege, and class.

j. The same tendencies that contribute to effective organization and bureaucracy, as discussed in (*a*), could change their import by being used as a general means of *dividing and conquering* and making the organization more entrenched and indispensable for the coordination of the parts. In the context of the work place, the ten tendencies can be used to 'deskill' a workforce and create factory discipline.[12]

k. Classification (*1*) and mold (*8*) especially can be used (unintentionally) to *obscure the mismatch* of territory and events by making people believe that the assignment of the particular tasks to the particular territories is indeed appropriate, when in fact the tasks are assigned to the wrong scale. An example of this would be assigning major responsibility for funding pollution abatement to local levels of government when in fact the sources of particular pollutants are not local.

l. Displacement (*5*) and territorial multiplication (*10*) could *direct attention away from causes of social conflict to conflicts among territories* themselves. Examples of this can be seen in the attention given to the urban crises and to the conflicts between the inner city versus the suburbs and the Snowbelt versus the Sunbelt, rather than to social-economic relationships causing the conflicts.

m. Molding (*8*) the geography of actions at various scales, coupled with enforcing long- and short-range planning responsibilities to corresponding levels of the hierarchy (*3*), gives organizations the opportunity to *obscure the geographic impact of an event*. This occurs by correctly specifying the geography at one scale, say the national, and not at the others, or by dividing a decision into parts, so that the initiation of an action (that may be irreversible) is considered in the context of the largest territory and the implementation of the action is left later to the smaller territories.[13] A combination of the two is found in the history of the United States policy regarding nuclear power. It was decided at the national level that a significant fraction of our electricity would be generated by nuclear power. This goal pertained to the nation as a whole and was well under way before the decisions to locate the plants were made at the local levels and before the decisions to dispose of waste were even contemplated.

n. The same tendencies that could help to make hierarchical organizational control effective (*1*, *2*, *3*, *6*, and *8* or *a* and *j*) could backfire, leading instead to a reduction of control and even to *secession*. Dividing, conquering, de-skilling, and making relationships impersonal may be nullified or offset by the potentials they have of creating disorganization, alienation, and hostility. In many cases the assembly line went too far in circumscribing and de-skilling. Workers have reacted to senseless assignments and to alienation with various degrees of resistance, and industry has recently begun to explore new kinds of organizations aimed at decreasing the territorial circumscription of workers at the lower levels of the hierarchy.[14] Moreover, those who resist circumscription can make use of the existing territories in various ways, as when prisoners literally take possession of cells and cell blocks, or as when political units secede. In cases where the seceding units take a territorial form, we would hypothesize that the reasons for employing territory would come from among the ten tendencies and their combinations.

These potentials, then, are not isolated and independent. The matrix in Figure 2.1, along with the preceding descriptions of the combinations, make it clear that some of the combinations use exactly the same tendencies as do others, but differ in the weights assigned to them and in the emphases placed on their connotations and normative meanings. For example, hierarchy and bureaucracy (*a*) and divide and conquer (*j*), and secession (*n*) all rely heavily on tendencies (*1*, *2*, *3*, *6*, and *8*) but they do so with different imports. Hierarchy and bureaucracy (*a*) can be thought of either as a benevolent or neutral organization using territory. Divide and conquer (*j*) emphasizes the negative aspects of (*1*, *2*, *3*, *6*, and *8*) and describes what might be thought of as a malevolent organization. Secession (*n*) describes the condition wherein an individual or group uses territorial tendencies (*1*, *2*, *3*, *6*, and *8*) to lessen or remove the authority of others. Similarly, obfuscation by assigning the wrong scale territory (*k*) is the malevolent side of mismatch and spillover (*g*). Social conflict obscured by territorial conflicts places a different emphasis on the same tendencies than does territoriality as an end (*h*). Obfuscation by stages (in terms of time and scale) (*m*) is the negative side of long- and short-range planning (*b*); inequalities (*l*) is the negative side of efficient supervision – span of control (*d*).

These fourteen combinations along with the ten primary tendencies are potential reasons/causes, consequences/effects of territoriality that are linked to our definition. The remainer of the book will illustrate that these delimit the domain of potential advantages at a precise and general enough level to be historically significant. Some combinations, such as divide and conquer, have been associated with territoriality before but most of the tendencies and combinations have not. This is unfortunate because they are necessary components in understanding how even the familiar territorial effects operate. We understand more of territoriality's role in dividing and conquering when we realize that territoriality allows the joint employment of the ten tendencies or that using these ten with slightly different emphases can help organizations become hierarchical and bureaucratic or can help create organization inefficiencies rather than help them to divide and conquer.

Some of the combinations can be collapsed into more general categories, as in joining all combinations that can be *obfuscatory* (*k*, *l*, and *m*) – these indeed form an important component of modernity under capitalism according to Marxist theory, as we shall see, and some can even be further subdivided. Certainly they all can be made narrower as when mismatch by spillover (*g*) is replaced by the narrower economic concept of externality, or divide and conquer (*j*) is replaced by the narrower example of nineteenth-century British colonial policy in Africa. But to do any of this runs the risk of being either too general or too specific. Again, that a definition and its entailments are at a proper level cannot be proven abstractly. We can only illustrate the utility of the theory by exploring case studies of territoriality.

Which potentials are more interrelated and what their interconnections look like constitute the theory's internal dynamics and structure. Figure 2.1 provides some suggestions about more or less likely interconnections: about how some potentialities can reinforce and some negate others. Overall there is the suggestion that territoriality can help increase the efficiency of an organization (whether it be a state, a business, or a church) up to a point, and that it can help shift an organization's goals from benign to malevolent. For instance, defining responsibilities territorially can be efficient, but it can also create inadvertent spillovers and mismatches when the territorial definition becomes a substitute for not knowing what it is that is being controlled. These inefficiencies can lead to the need for more hierarchy and larger territories to coordinate the spillovers and mismatches. But eventually central control will be impaired. This could result in local levels having greater autonomy *de facto* if not *de jure*. Defining responsibility by area can also be used intentionally to obscure or disguise processes, increase the advantages of those in control, and shift the organization from benign to malevolent.

Some of these suggestions are illustrated in Figure 2.2. This diagram begins with the assumption (illustrated by the straight path to 'a') that the original goals of the organization are benign or neutral and that the institution draws from among the tendencies of territoriality to increase its hierarchical control. But some of the internal loops and tipping points suggested above may take hold and make the organization inefficient. This

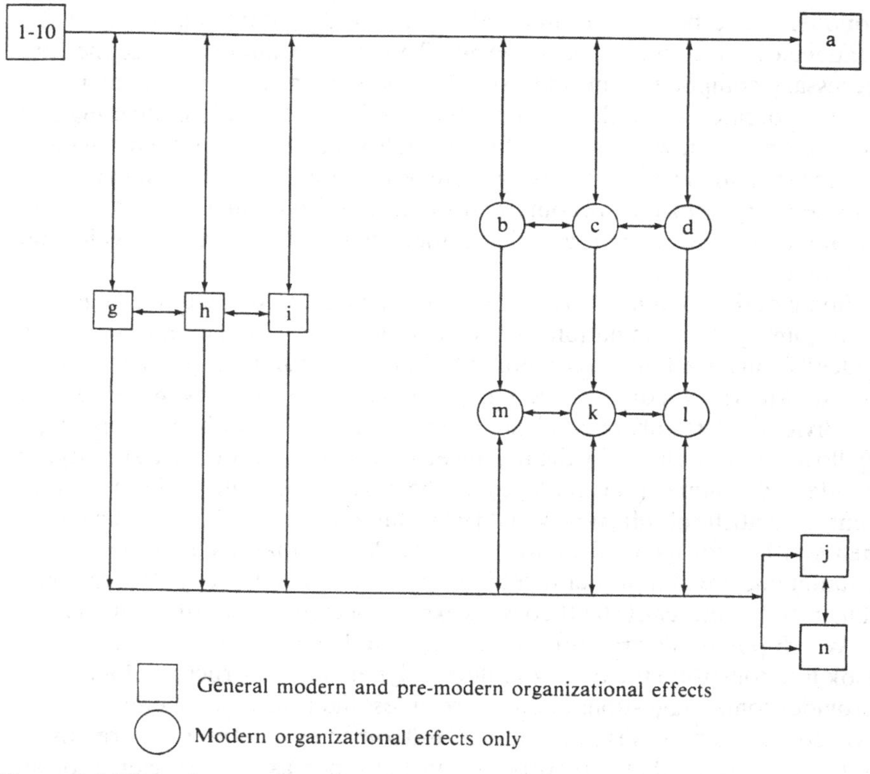

Figure 2.2 Flows, loops, and tipping points

can increase local autonomy and even fragment the organization, or it can push the organization away from neutrality to a malevolent state of dividing and conquering those that it is controlling. The circles point to loops that are especially prevalent in modern society.

These are only a few of the many possibilities arising from the interconnections among the potentials. In order to follow them further, the theory must become more explicit about the types of social context that will be employing particular potentialities of territoriality. Indeed, it must be remembered that social context has never been completely ignored. To a very general degree it is woven into the definition of territoriality itself. We have only pushed it into the background, more so with the tendencies and less so with the combinations. The logic of territoriality can carry the discussion still further, but only by combining it with more and more explicit types of social contexts which can be expected to utilize territoriality. Social context must now come into the foreground in order to fill in more of the internal structure of the theory, just as the periodic table of elements must be available for the atomic structure of an atom to make sense.

Theory: part II

Bondings: history and theory

Social science is acquainted with numerous types and modes of societies. To focus the discussion we will concentrate on social models of Weber and Marx. In addition to being enormously influential, these models have addressed a broad range of social organizations and have a great deal to do with territorial structure. By no means are they the only models to which the theory can be combined.

Weber Two facets especially to Weber's work have a bearing on our discussion. The first considers the internal dynamics of organizations and especially of bureaucracies, and the second addresses the historical–social context in which certain organizations are more or less likely to occur.

Taking the second first, we note that Weber refers to three general or ideal types of organizations: charismatic, traditional, and bureaucratic. The first is not necessarily linked to any period or type of society.[15] Its followers and leaders form a loose organization. There are few if any officers, rules of procedure, and clear hierarchies. But as the group persists, and especially as the question of succession arises, charisma becomes 'routinized.' It gives way to one or the other of two more formal types of organizations: the traditional and the bureaucratic.

As the name implies, traditional organizations are found primarily in pre-modern societies or civilizations containing social classes and complex divisions of labor. These organizations rely on traditional modes of conduct and problem solving. Often, the leadership is drawn from a specific clan, family, or circle of friends. Justification for authority is based on custom. Hierarchy can be very well developed and complex, but a person's ability and personality may change the power and scope of his appointment. Legitimacy of authority is not drawn from holding an office proper but from being connected to traditional positions of leadership. Traditional organizations occur throughout pre-modern civilizations and they characterize organizations when charisma becomes routinized. Many scholars have called such traditional hierarchies bureaucracies[16] but Weber reserves the term for organizational features found primarily in modern societies which include capitalistic and socialistic economies. We will follow the practice of calling all such organizations bureaucratic, but point to the degree to which they contain modern features such as the ones Weber notes. The routinization of charisma in modern society according to Weber would normally lead to bureaucratic organization. Bureaucracies, in Weber's terms, are characterized by formal lines of communication, clear hierarchy and definitions of authority, and impersonal relations. These make hierarchical organizations modern.

More specifically, Weber argues that:

> 1) the individual office holders of modern bureaucracies are personally free and subject to authority only with respect to their impersonal official obligations. 2) The bureaucracies themselves are organized in clearly defined hierarchies of offices. 3) Each office has a clearly defined sphere of competence in the legal sense. 4) The office is filled by a free contractual relationship. 5) Candidates are selected on the basis of technical qualifications . . . [often] by examination or diplomas certifying technical training or both. Candidates are appointed, not elected. 6) They are remunerated by fixed salaries in money . . . The salary is primarily graded according to rank in the hierarchy . . . 7) The office is treated as the sole or at least the primary occupation of the incumbent. 8) It constitutes a career. There is a system of 'promotion' according to seniority or to achievement or both. Promotion is dependent on the judgment of superiors. 9) The official works entirely separated from ownership of the means of administration. 10) He is subject to strict and systematic discipline and control in the conduct of the office.[17]

Conversely, hierarchical (bureaucratic) organizations that are not modern tend not to have: a) a clearly defined sphere of competence subject to impersonal rules; b) a rational ordering of relations of superiority and inferiority; c) a regular system of appointments and promotions on the basis of a free contract or technical training; and d) fixed salaries. Note that impersonality is a major undercurrent in the list. The higher the degree of impersonality, the more modern the bureaucracy.

Much recent research on organizational structure has consolidated and extended Weber's components and has contained specific suggestions about their interconnections. The objects of study have for the most part been *twentieth*-century Western industrial organizations but several of the variables and their supposed interconnections can serve as guides to an analysis of pre-modern institutions.

These recent works, as with Weber's, point to the significance of impersonality and impartiality within modern organizational and bureaucratic structures, and also suggest the following as important facets of organizations:

Specialization – which refers to the division of labor;
Standardization – which refers to the extent of procedural regularity in the organization;
Formalization – which refers to the use of documentation for job definition and communication;
Centralization – which refers to the locus of authority in the organization;
Configuration – which refers to the shape of authority and hierarchy and can often be summarized by span of control.

These are of course very general characteristics. More specific meanings differ considerably from study to study. Yet there is consensus that specialization, standardization, and formalization are strongly interrelated and are connected to the hierarchical structure of organizations and also to technology.[18]

Little modification has been made to the historical facet of Weber's formulation, except that, as we noted, others have used the term bureaucracy more generally to describe the hierarchies in traditional organizations and have found within them some examples of modern bureaucratic facets, such as impersonal relations. Rather, most amendments have been made to the first aspect of Weber's formulation, the description of processes occurring within modern organizations and bureaucracies, and here two avenues of research have a bearing on territoriality.

First, as we have just described, recent work on organizational structures has introduced such facets as standardization, formalization, centralization, and configuration, to consolidate components of Weber's model. Second, research on organizations has explicitly addressed the normative implications of bureaucracy. Weber saw the bureaucratic form as potentially the most rational and efficient. He recognized some of its negative features, such as its tendency to make relationships too uniform and impersonal, which would cause the organization to dissolve or split apart and could create opportunities for charismatic leaders to form new ones. But he was most impressed with bureaucracy's positive potentials of rationality and efficiency. Overall he presented the bureaucracy as an instrument with the potential to do good.

Bureaucracy's negative side was investigated and elaborated more fully by Weber's successors, especially Michels and Merton. Michels examined German socialist organizations and found that, despite their idealistic and egalitarian beginnings, these organizations became increasingly institutionalized, authoritarian, and hierarchically rigid; and the officials became more interested in perpetuating themselves and their offices than in their commitment to the original goals of the organization.[19] This trend he attributed to bureaucracies in general and called it the 'iron law of oligarchy.' Merton disclosed another malevolent side to bureaucracy. An emphasis on strict formal procedures, discipline, and rules, he argued, leaves officials with the view that adherence to formal procedures is an end in itself.[20] This Merton called 'displacement.'

Many other studies of bureaucracy's problems can be cited, and their collective import is that although Weber's characterization was not wrong, there is more to the internal dynamics of bureaucracies which often leads them away from efficiency and benign or neutral effects. Granted, then, that organizations are dynamic, that modern society has complex hierarchical organizations with particular characteristics which Weber calls bureaucratic, and that traditional societies possess traditional though often complex hierarchies with few modern bureaucratic characteristics, how can this be linked to territoriality?

The union occurs because many of these dynamics are mirrored in the logic of territoriality and because both traditional and modern organizations have employed territoriality as integral parts of their structures. Joining research on modern facets of organization with territoriality can lead to the following expectations. In very general terms, the theory would suggest that in both traditional and modern society territoriality could increase organizational efficiency, centralization, and span of control, but again up to a point. The theory also anticipates, as in Figure 2.2, that tipping points can be reached making it possible for territoriality to weaken an institution. The territorial units can secede or become captured by another organization. The process may be subtle as when territorial units engender bureaucratic inefficiencies and become ends in themselves. If we focus especially on modern facets of bureaucracy, we can expect that in modern society, but also to some degree in pre-modern ones (as will be shown in our discussion of the Catholic Church), territoriality's facility in providing ease of classification, communication, and control could also increase specialization, standardization, and formalization up to a point. The expression 'up to a point' must be emphasized again because the society in which these organizations occur has much to do with the specifics of these relationships and because the critical tipping points in the internal dynamics of the theory can again eventually come to bear to counter some of these effects. These tipping points are the territorial equivalents to bureaucracy's conservative and oligarchical effects.

The theory's internal logic can be refined to yield even more specific relationships when the type of organization using territoriality is more clearly defined. As we shall see in Chapter 6, for modern centralized bureaucratic organizations like the military, the school, and the factory, specifiable quantitative relationships can be expected to hold among degrees of territoriality, span of control, hierarchy, task complexity, and technology.

Whereas many of the relationships between territoriality and complex hierarchical social divisions of labor can be present to varying degrees in practically any organization, we should not lose sight of the fact that some would predominate in modern society. This means that many of territoriality's uses within an organization depend on the society in which the organization occurs. Governments of empires and modern states have territorially subdivided their domains because territoriality can provide these organizations with advantages. But just as there are differences between traditional organizations and modern bureaucracies, so too are there differences in the territorial effects they employ. Comparing the dynamics of these types of organizations with the potential dynamics of territoriality can help specify the conditions of each. For instance, traditional organizations, unlike modern bureaucracies, would not be expected to emphasize territoriality's effects of creating impersonal relationships and of conceptually emptying space.

Still there are the fascinating and important cases of pre-modern organizations containing some of these modern effects. As we shall see, the Catholic Church is a case in point, but so are the Chinese Mandarin system of selecting officials and the English feudal system of King's Courts. These systems possessed territorial devices to help keep relationships impersonal. One such device was to rotate officials from one territory to another, or at least not to assign an official to his native region.

Marx A second major social theory that could be linked fruitfully with territoriality is Marxism. Marx did not examine the possibility of bureaucratic dynamics as an independent phenomenon. Rather his writings discuss bureaucracy as an institution to be manipulated by class power. This is because Marx taught that the social division of labor, as manifested in ranks, specializations, and roles, is determined by the economic division of labor. The twists and turns of bureaucracy are linked to the development of economic classes and their interrelationships. Once communism removes class conflict, the state, as an agent of oppression, would wither away. Marx did not directly address the question of whether bureaucracy would also wither away along with the state, but in his early critique of Hegel he sees socialism as simplifying the bureaucratization of the state.[21] Recently Marxists have recognized that bureaucratization is a force to be reckoned with in socialist countries, if not in the utopian world of communism. The Soviet bureaucracies have internal dynamics and contradictions of their own. The oligarchical tendencies of government bureaucracy, for instance, can create the equivalent of class structure and interests, and their forms, imports, and dynamics are effected by their social-historical contexts.[22] This literature, then, could add further specifications to the directions and imports of the dynamics within bureaucracy.

More directly to our purposes is that the Marxists' theory of class conflict in capitalism, when applied to territoriality, would single out the *obfuscatory* combinations of territoriality (*k*, *l*, and *m*) as the most important in the later stages of capitalism.[23] The obfuscatory combinations would be expected because of the general tendency of capitalism to disguise class conflict and because of the peculiar position of the state *vis-à-vis* labor and capital which has a particularly important bearing on the theory of the state. On the one hand the state tries to maintain capitalism, and on the other it must contain or reduce class conflict, claiming to be the champion of the people and a vehicle for providing public goods. This dual role means that the sources and forms of power must often be disguised and the obfuscatory tendencies of territoriality could help do this. Territorial obfuscation need not be applied only at the state or local-state level. It could appear as well in the work place, the school, and in the realms of consumption.

Moreover, Marxist theory, in conjunction with a general analysis of modernity, points to the present and the recent past as the times to expect the most intense and frequent occurrence of *emptiable space* (*e*). This is because capitalism reinforces the view of space as a framework for the location and distribution of events. Capitalism helps turn place into commodities. It helps us see the earth's surface as a spatial framework in which events are contingently and temporally located. Capitalism's need for capital accumulation and growth makes change paramount and, geographically, change means a fluid relationship between things and space. The future is conceived of, and future actions produce, continual alterations of geographical relationships. Territoriality then becomes the mold for both filling space and defining and holding a space empty.

Weberians, as well as Marxists, would point to the fact that pre-modern civilizations may have differed in their uses of territoriality but that the differences in uses among them are in several respects not as great as the differences between their uses and those of modern society. They would also agree that only one other comparable historical watershed has occurred in territorial use, and that took place in the transition between primitive society and civilization. Marx, and Engels especially, characterize the primitive as essentially different from other pre-capitalist modes.[24] To them, the primitive means small-scale egalitarian society with few if any institutions of oppression. The primitive's use of territoriality would be quite different from that found in civilizations, whether pre-capitalist or capitalist. For instance, in primitive society one would not expect to find frequent or intense use of territoriality to form impersonal relations (*6*), to mold (*8*), conceptually to empty place (*9*), or to multiply territories (*10*), and one would not expect to find most of the combinations, especially territorial definition of social relationships (*c*).

There is much more that could be said about the links between the theory of territoriality and Marxist, Weberian, or other theories of power and organization. More specific connections can and will be made later in the book and examined in concrete historical cases. In discussing modernity in subsequent chapters we will be considering the interpretations of territoriality by neo-Smithians and neo-Keynesians as well as by Weberians and Marxians. But this sketch is sufficient to point to the possible areas upon which a history of territoriality should concentrate.

Figure 2.3 summarizes the principal relationships we mentioned between uses of territoriality and their association with those social-historical contexts suggested especially by Marx, Weber, and a general understanding of history. It emphasizes the broad connection between territoriality and

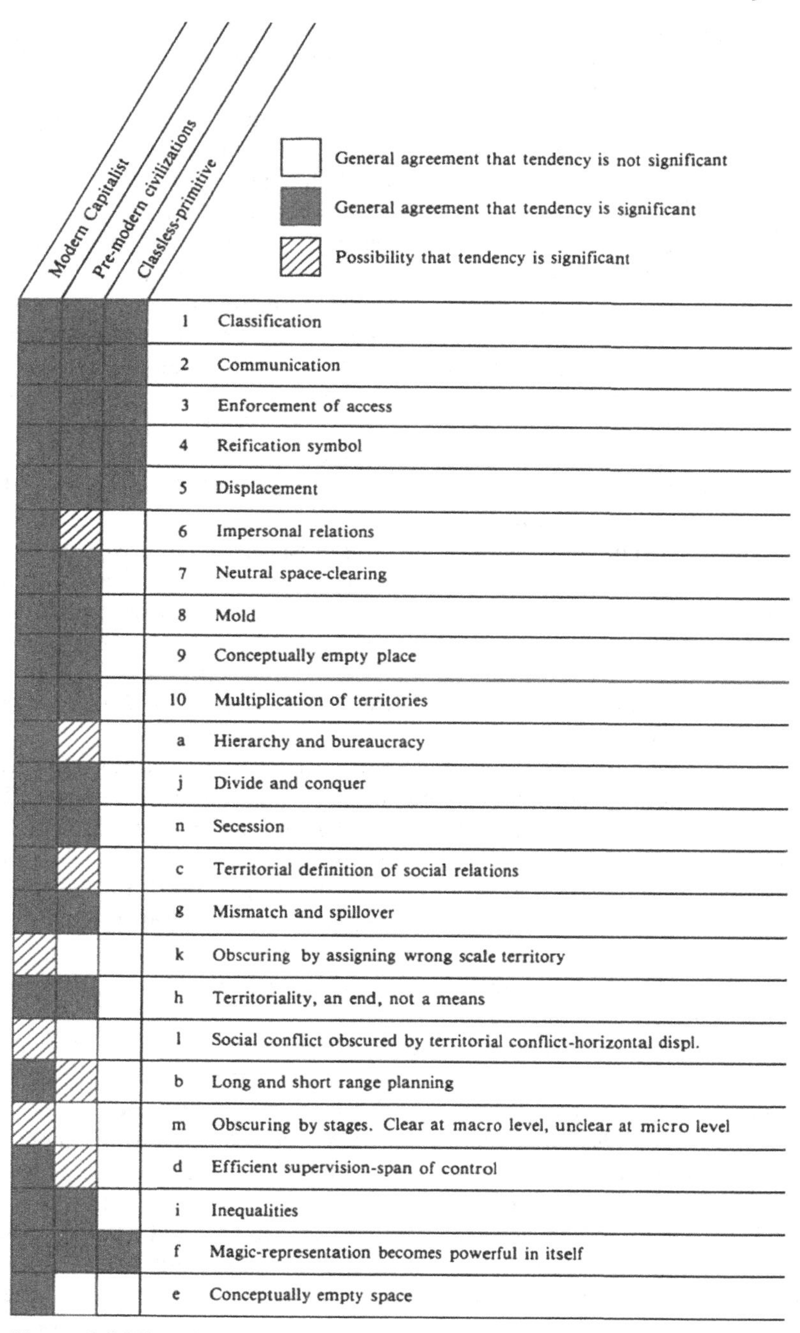

Figure 2.3 Historical links

change in political economy. It thus considers an economic division of labor to be a primary factor in determining who controls whom and for what purposes. This does not mean of course that territoriality is unaffected by other economic as well as non-economic factors – indeed Figure 2.2 suggests how hierarchical organization is one such influence on territoriality. Rather it proposes that the most general territorial changes can be associated with changes in political economy. The two critical historical periods indicated in Figure 2.3 are the rise of civilizations and the rise of capitalism and modernity. (Although capitalism and modernity are not synonymous – the latter includes cultural and ideological components that are not reducible to economic terms – capitalism is an historically crucial element of modernism and for the sake of brevity the two terms will sometimes be used interchangeably. So called 'socialist' countries such as the Soviet Union are also modern, but as there is little consensus about their political economic form, their use of territoriality will not be discussed separately.)

In the rise of civilizations, the most important novel effects of territoriality are its uses in governing others, in defining social relationships, in dividing, subduing, and organizing populations. The most important novel territorial effects accompanying the rise of capitalism are its uses in conceptually emptying space, in creating modern bureaucracies, and in obscuring sources of power. These are sketches of the overriding connections between historical organization and territorial functions, and, along with Figure 2.2 and the general suggestions contained in Figure 1.2, will serve as the organizing models for the rest of the book.

Figure 2.3 summarizes ideas about the past, but so too does all written history, as opposed perhaps to what actually happened in the past. The succeeding chapters will attempt to show that the historical record (including what others than Marx and Weber have said about the past) tends to bear out these hypothesized associations. I use the words 'tends to' for several reasons. The span of time and space to be covered is vast. Only the highlights can be sketched which means judiciously selecting cases and sampling the vast quantity of secondary sources. Historical interpretation is in continuous flux; alternative views abound for practically any period as do debates about the period's importance and duration. What capitalism is and when it became important is scarcely resolved. The same is true for other periods and social organizations. Although what Marx and Weber offer are general models (and models, especially social ones, are only partial and approximate representations of reality), large numbers of more detailed historical studies can be found which tend to accept their general social typologies. Evidence from these detailed works will be used in the next chapter to flesh out the relationships suggested in Figure 2.3; and especially to highlight the two overall transitions – from primitive to civilized, and from pre-capitalist to capitalist. These changes will also be described in conjunction with changes in the conceptions and uses of space and time. Chapter 3, then, is an overview of the history of 'Territoriality, space, and time.'

Figure 2.3 emphasizes the economic forces behind social-territorial relationships. But it also suggests that several territorial effects, such as defining things by area, and enforcing access, can occur in any society, whereas others, such as the use of territoriality to make relationships impersonal and to increase span of control, can be found in all civilizations, but more so in modern than in pre-modern ones. In other words, institutions with modern characteristics can exist within pre-modern societies and vice versa. Moreover, what is not described in Figure 2.3, but is implied in the general matrix of Figure 2.1 and is illustrated in Figure 2.2, is that the relationships among people and territory can undergo important changes while the society in which they occur does not. Even though the political economy of a society (at least at the level of generalization described in Figure 2.3) may remain the same, the territorial effects between people and within an organization, as illustrated by Figure 2.2, may have a dynamic of their own. Chapter 4, on the Church, will explore these possibilities by examining the internal territorial dynamics of Church organization and its relationship to political economic change.

The Church is one of the most enduring and best-documented examples of an institution using territoriality as an integral part of its organization. The parish, the diocese, the archdiocese, and also the architectural partitioning of church buildings, clearly reveal the Church's reliance on territoriality. These aspects of Church organization were developed differently during three major historical periods, the Classical, the Feudal, and the Modern. Yet during the last period, the Church's internal dynamics seem to have resisted many of the external political-economic changes accompanying modernization. In the Roman and Feudal periods, the territorial effects of the Church were far more modern than was the case with other institutions of the time, but by the end of the Middle Ages the same effects served to separate the Church from society and insulate it from social change, so that since the Reformation the Church stands more as an example of an archaic than a modern organization. This mixture of change and persistence, of internal dynamics and resistance to change, makes the Church an important case study of the interconnections among hierarchy, bureaucracy, and territoriality, from pre-modern times to the present.

Chapter 5,'The American territorial system,' will use the political formation of North America, from the sixteenth century to the present, to concentrate again on the themes of modern territoriality use, especially their role in creating a concept of emptiable space, in facilitating bureaucracy, and in obscuring sources of power. Chapter 6, 'The work place,' will concentrate on how these same modern uses of territoriality have developed in the last 300 years at the local and architectural levels.

3

Historical models: territoriality, space, and time

For humans, territoriality is a strategy to affect, influence, and control. It is always used in conjunction with non-territorial spatial strategies. Both the selection of territoriality and the effect it has depends on social context: on how space in general is used and conceived as well as on who is controlling whom and for what purposes. This means that the history of territoriality is closely bound to the history of space, time, and social organization.[1] Before exploring these interconnections, and especially territoriality within the two historical watersheds – the rise of civilization and the rise of capitalism and the modern world – it would be well to point out some general historical trends of territoriality.

Trends and complexities

First, there is strong indirect evidence that the total number of *autonomous* territorial units in the world has declined enormously from prehistoric times to the present. The decline has not been uninterrupted, but, since the rise of civilizations, there is no doubt about its general directions. The entire world population during the upper paleolithic age may have been no more than 3 million. Yet mankind was divided among 100,000 or more small independent entities.[2] By observing contemporary pre-literate hunting–gathering societies and reflecting upon the reasons why such groups have been territorial in the past, we can surmise that, although each of these pre-historic peoples occupied geographic area, not all of them may have made use of territoriality and many may have done so only intermittently. Still there were enough to place the number of territorially autonomous units in the tens of thousands. This is a very large number when compared with the 150 nation states of the 1980s, many of which are only somewhat independent or autonomous.

Second, since pre-historic times the size of these autonomous territorial units has increased, from the circumscribed hunting–gathering collecting areas or even smaller agricultural areas of villages and households, to enormous claims of empires and nation states. (As we noted with the Chippewa, the few hunting–gathering societies that had large ranges could not have been territorial over their entire range. Indeed a reason for their vast and often non-territorial domains may have been the scarcity and unpredictability of food and other resources. Even if they could enforce territoriality this unpredictability would have made territoriality a poor strategy.)

Third, is that these ever fewer and larger autonomous units have become increasingly subdivided and fragmented into varied territorial sub-units. These form territorial hierarchies within the society. It is through this subdivision that the total number of territories of all but the autonomous type has increased. Ancient empires were subdivided into a multiplicity of smaller layers of jurisdictions. So too were places of production and consumption. The creation of large building complexes, subdivided into courtyards and rooms, indicates the layouts of the territories at the smallest geographical levels. The modern nation state has unprecedented territorial levels for every realm of life. Political units are hierarchically ordered from nation state to local wards and even into single purpose districts; and offices, factories, and houses have become honeycombed with hierarchies of their own.

Understanding these overall patterns and the numerous effects they encompass requires unpacking the types of territories and the historical and social contexts in which they occur – especially knowing who is influencing and controlling whom and why. The theory of territoriality points out that societies that do not have formal hierarchies, economic classes, and other types of institutionalized differences would use territoriality in a different way than those that do. Historians and pre-historians suggest that relatively non-hierarchical societies (similar to the aboriginal Chippewa) were common before the rise of civilization and that some have persisted to varying degrees in remote parts of the world to this day. These have been called for so long 'primitive' societies that we also will employ the term despite the fact that it has come to have a negative connotation in that its use is thought to indicate a condescension on the part of the presumably non-primitive person using it. However, the original meaning of the term is in no sense pejorative. It means primary or original in time and even in rank.[3] Our use of the term is intended to convey this original non-pejorative meaning. We are interested in the earliest, original, or primitive types of societies and their uses of space and whether these uses conform to those outlined in the discussion in Chapter 2.

A further point about the term primitive is that it also implies that humans have progressed through at least two stages – from primitive to non-primitive or civilized. But what exactly do the ideas of stage and progress mean? Some would have it that stage is a necessary sequence in which the former society in some sense causes or gives rise to the latter, and that stage is also a necessary progression towards some higher or better form. Such meanings of stages are impossible to support in the case of human history and are certainly not what is intended by our interests in the primitive, in the transitions from primitive to civilized, or for that matter in the transitions from feudalism to capitalism. Rather our interest in particular forms of social organization, including the primitive, is intended to point to the fact that very different types of social organizations have existed, that some have predominated only during particular periods, and that some helped occasion others.[4]

These changes and sequences, though, must be divorced from the notions that they were somehow inevitable, necessarily for the better, and in one direction only. Rather the idea of change must allow for the fact that earlier forms of social organizations are not entirely replaced by newer forms, and that newer forms may have some earlier representations. The idea of change then must be sensitive to the fact that past organizations persist, though in altered forms, within present structures, and that, though a particular structure may predominate, a society usually contains other forms intermixed.

With reference to the primitive, we can say that before 7,000 years ago there were no societies whose characteristics were dominated by those that are associated with civilization. But shortly after 7,000 years ago we find that several societies were coming to possess important characteristics associated with civilization. This does not mean that primitive societies did not contain traces of attributes associated with civilizations or that civilizations did not contain attributes of primitive societies. Indeed boundaries of early civilizations encompassed primitive communities, and primitive communities continued to persist beyond civilizations' frontiers. However, we must assume that those that persisted even beyond the domains of empires were eventually, though often indirectly, altered by the presence of civilizations. Perhaps their customs and material cultures were changed, or perhaps they were simply forced to move to new regions because of the indirect population pressures exerted by an expanding civilization. It is this intermixture of old and new, not only as it pertained to the rise of civilizations, but as it pertained to the rise of capitalism, that is intended by our discussion of historical transitions and changes.

Returning to the characterization of the primitive, it is important for us to construct a minimal description of this society which best conforms to the evidence and which addresses issues about its use of space and territoriality. But evidence is not easily had. Knowledge about the primitive is indirect and comes from two different sources. First is the archeological record. Second are written observations throughout history of societies having comparable technologies to those that were to be unearthed by archeologists. Yet these written records are about societies that undoubtedly had changed and that were influenced by contacts with civilizations. Nevertheless, their technologies, social organizations, and systems of beliefs are on the one hand quite different from those of civilizations, and on the other share strong resemblances to what we know of the primitive societies from archeological remains. Abstracting and combining the two kinds of evidence presents a composite picture which, with proper precautions, will serve as a characterization of primitive societies.

Primitive political economies

Household economy

Since territoriality is a geographical form of power and its import depends on who is controlling whom and for what purposes, we should begin our explorations of territoriality in primitive societies with a general picture of how they use power and authority. This is best achieved by examining their political economy.

Overall, primitive societies are less complex than civilizations; they have less division of labor, internal specialization, fewer numbers of people, and contain smaller geographic areas. Their livelihood can be obtained through any combination of gathering, hunting, foraging, herding, and agriculture. In virtually every primitive group the household is the basic unit of production and consumption. This unit may be composed of a primary or an 'extended' family. An amalgam of households (whether related or not) can be called a band.[5]

In hunting, gathering, and foraging societies, bands are often seasonal. Households move their locations according to the distribution of resources and often join together in bands when it is time for cooperative efforts. The sizes of bands vary, but rarely do they approach the size of even a modest modern town. Of the extant hunters and gatherers, perhaps the Eskimos have the largest villages, numbering in places of good hunting several hundred inhabitants. Clans and tribes are more complex than bands and may be held together through kinship and lineages, real and fictitious. Indeed, the term 'tribe' refers to a range of organizational types, many of which have been a response of primitive society to pressures from civilization.[6] For primitive society the social forces binding individuals to larger groups tend to be weaker as the size or level of complexity of the group increases. It is strongest within the household and weakest within the tribe, if and when it exists.[7] The general term 'community' will be used to embrace the range of social units to which the household may belong.

The nature of the primitive economy lies in the interrelationships among

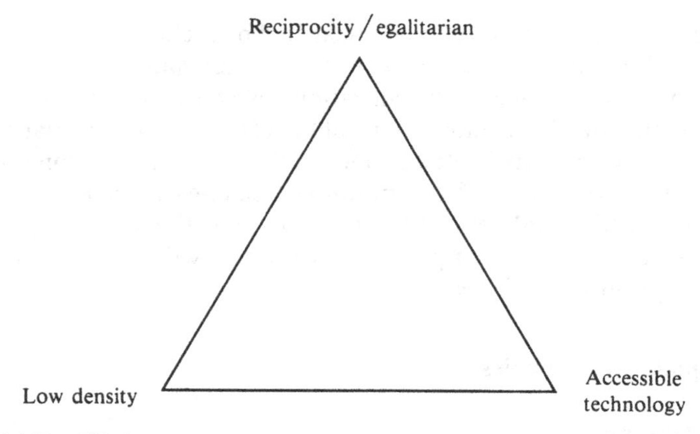

Figure 3.1 Equilibrium in primitive economies: technology density, reciprocity

their small size, their accessible technology, and their communal–egalitarian values. Figure 3.1 illustrates how each of these reinforces the other.[8] By most accounts primitive society lacks economic classes, inherited 'wealth,' and legal, political, or administrative institutions or apparati of state apart from and above the people. Primitive society is participatory. Conflicts within the communities are not directed against institutions or corporate entities, but against specific individuals. Leadership is by consensus, and leaders are those who rise to the occasion. There may be a leader of the hunt, another for spiritual matters, another who is a wise man, and so on. These roles are not coercive. Nor are they inherited. The primary institutionalized uses of power that may be restrictive – non-egalitarian – are those that would assign responsibilities on the basis of age or sex. This consensual basis of politics is closely related to the egalitarian nature of the economic system. One frequently hears that no one starves in a primitive community unless everyone starves. Often a good leader of the hunt gives away his kill and then receives a (modest) portion back for himself and his dependants.[9]

Sharing takes the form of reciprocity and is reinforced as Figure 3.1 suggests by the small size of the communities.[10] If a strong man begins to wield power unjustly, or unwisely, those who are discontented or maligned might leave to form their own community or join another. If this is not possible, community could ostracize the bully. Because these are small communities, the effects of one's actions on others can be seen quickly by all. This intimacy and the economic interdependence of the community members makes mutual aid and cooperation all important.

Again as Figure 3.1 indicates, economic inequality is made even more unlikely by having an accessible technology. Tools may be very complex and intricate, but knowledge about how to produce them is widely held. Although some might be able to make them better than others, this does not make it impossible for others to produce them. Knowledge of tool making could not be monopolized. Nor could someone within the society monopolize the game, the fish, the berries, or the nuts. Even if the society is agricultural, there is rarely a shortage of arable land, and so again monopolization of plots from within the community is unlikely.

This does not mean, however, that the community cannot assign different plots of land to be used by different households. This might frequently have been the case, but the land would remain the community's. Assignments would be based on household need and would be periodically re-evaluated. Land itself may not always be the primary focus of value. Often distinctions are made between land and its produce. The land itself may not be assigned to anyone in particular, rather a person or a family may be given something like the right to harvest fruit-bearing trees.

Family, kin, and ritualized friendships provide the complex channels of reciprocity through which labor, resources, and products flow to equalize discrepancies and to share in times of emergencies. Could not these reciprocal familial obligations contain unequal sets of relationships? They could. Age and sex as we noted may create inequalities and there are likely to be others that have to do with ability and personality. But these are not going to get out of control and allow individuals and families to monopolize power and resources because of the constraints placed on inequality by accessible technology and low population density. In short, under these conditions reciprocity means an equal return of obligation.

We should stress that the egalitarianism of these communities applies primarily to the circulation and consumption of necessities, and that numerous types of inequalities exist in other realms. People differ in talents and abilities, and primitive peoples also possess personal or 'private' property. They may own their 'own' weapons, their own leantos, their own teepees and their own clothing, and in some cases even their own legends. But these differences in abilities and possessions do not prevent others from gaining a livelihood. Also, although these communities are small, personal, and often closely knit, they are not necessarily harmonious. Conflicts, even violent ones, are documented in ethnological observations and are also part of many of these people's own legends.[11]

The small size of these communities and their lack of specialization and division of labor affects not only their political economy, but also their relationships to nature. It makes their contact with their natural surroundings appear close and intimate. This is frequently expressed in their use of family as an analogy for the relationships between society and world. This analogical extension of the family contributes to the general primitive view of the unity of nature. It presents nature as personal and vital. It creates a sense of reciprocity between man and nature, and leads to important conceptions of space, place, time, and territory.[12]

Territoriality

The connection between a primitive people and the place they occupy becomes extremely close (leaving aside for the moment whether or not the place is territorial) not only because of familiarity and dependence but also because the people come to think of themselves and place as organically and even spiritually linked. Their geographic domain may be either the entire area they occupy or only special and localized places. In either case, the land is often believed to be inhabited by the spirits of the ancestors and their geographic place in the world may have been given to them by their gods. In Australia, each totemic group is associated with a place from which the totemic ancestor is supposed to have emerged. When a person dies, his spirit returns to the place of his totemic origin. The Northern Aranda aborigine of Australia

> clings to his native soil with every fibre of his being. He will always speak of his own 'birthplace' with love and reverence. Today, tears will come into his eyes when he mentions an ancestral home site which has been sometimes unwittingly desecrated by the white usurpers of his group territory . . . Mountains and creeks and springs and water-holes are, to him, not merely interesting or beautiful scenic features . . .; they are the handiwork of ancestors from whom he himself has descended. He sees recorded on the surrounding landscape the ancient story of the lives and the deeds of the important beings whom he reveres; beings who for a brief space may take on human shape once more, beings, many of whom he has known in his own experience as his fathers and grandfathers and brothers, and as his mothers and sisters. The whole countryside is his living, age-old family tree. The story of his own totemic ancestor is to the native the account of his own doing at the beginning of time, at the dim dawn of life, when the world as he knows it now was being shaped and molded by all-powerful hands.[13]

Prominent landscape forms are often the ones incorporated into the myths which help to make the sense of space alive. According to Penobscot Indian lore, most of the landscape is a result of the peregrinations of the mythical personage Gluskabe. Gluskabe's snowshoe tracks are still impressed on the rocks near Mila, Maine; a twenty-five foot long rock near Castine is his overturned canoe; the rocks leading from it are his footprints; and Kineo mountain is his overturned cooking pot. A place on the earth in many creation myths was given to a people specifically by the gods. The Pawnee, for instance, believed that they were guided from within the earth to their present place by Mother Corn.[14]

Belief in the inhabitation of the land by the spirits of ancestors and in the mythical bestowal of the land to the people is one reason why it is likely that if the community were to assert territorial control over the land, they would be doing so through a social definition of territory. But before we explore these points further, we need to determine what conditions might make primitive communities use territoriality, if at all.

The theory suggests that territoriality can be expected simply because it can be an efficient device for establishing differential access when the resources to be controlled occur relatively predictably and densely in space and time. Clearing, planting, weeding, and harvesting are all relatively predictable, dense, and stable activities in space and time, and hence agricultural communities, if for no other reason than keeping animals out and preventing people from getting in each other's way, can be expected to use territoriality. The need for territoriality at the community level will increase if there is also competition for land from outside. The same logic applies to a territorial strategy for non-agriculturalists – hunters, gatherers, collectors. They can be expected to use territoriality if the resources they need are relatively predictable in space and time.

These expectations are borne out in the literature. Agricultural communities tend to be territorial as a whole and tend to subdivide land within the community, and these tendencies increase as population densities increase. Territoriality in hunting–gathering societies depends on the distribution of resources over space and time. Two socially–culturally similar societies, as are for instance the Utes and the Paiutes, differ in their degrees of territoriality because of the differences in the distribution of their resources in space and time.[15]

Efficiency of access is a reason for primitive society to employ a territorial strategy. But other closely related reasons can also be found in the evidence. For instance the ease by which a territorial boundary communicates possession and control can be of particular advantage in pre-literate societies. Also, to the degree that magic and ritual are extensive components of primitive communities, the reification and displacement effects of territoriality can be used to advantage in demarcating sacred places and areas of taboo. Combined, these effects make the sacred visible and the visible sacred.

But other potentialities such as creating impersonal relations, increasing span of control, effecting long- and short-range planning are simply not of use to a society that is not large and has no hierarchy and bureaucracy. And most importantly, a primitive society would most likely not have to define itself territorially. Recall that a social *vis-à-vis* a territorial definition of social relationships is always a matter of degree. A social definition means that to have access to the land one must be a member of the society, which in primitive society means participating in the spiritual history of the group. For example, in Bakongo tradition

> the ownership of the soil is collective, but this concept is very complex. It is the clan or family which owns the soils but the clan or family is not composed only of the living, but also, and primarily, of the dead; that is the *Bakulu*. The *Bakulu* are not all the dead of the clan; they are only its righteous ancestors, those who are leading a successful life in their villages under the earth. The members of the clan who do not

uphold the laws of the clan . . . are excluded from their society. It is the *Bakulu* who have acquired the clan's domain with its forests and rivers, its ponds and its springs; it is they who have been buried in this land. They continue to rule the land. They often return to their springs and rivers and ponds. The wild beasts of the bush and the forest are their goats, the birds are their poultry. It is they who 'give' the edible caterpillars of the trees, the fish of the rivers, the wine of the palm trees, the crops of the field. The members of the clan who are living on the soil can cultivate, harvest, hunt, fish; they make use of the ancestral domain, but it is the dead who remain its guardians. The clan and the soil it occupies constitute an indivisible thing, and the whole is under the rule of the *Bakulu*. It follows that the total alienation of the land or a part of it is something contrary to Bakongo mentality.[16]

A social definition of territory is found in all societies, but it constitutes the primary social sense of territory in primitive societies. The territorial advantage of classification by area and not by type or kind – which is a principal factor in a territorial definition of social relations – is unnecessary for a primitive community to use to define its own membership. The small size and close association of peoples in a primitive community means that its members are familiar enough with one another not to require a territory to define who is, and who is not, one of them. By the same token, territoriality is not needed to identify strangers or non-members. They are simply those who have no relationship to the members of the society.

One can suppose that a territorial definition of community could be helpful in those primitive groups or bands that allow relatively free entry and exit of its members.[17] In these cases simply establishing residence within a group's boundaries would make one a member of the community and allow one access to its resources and this would mean that such a community would then be relying on territoriality to define itself. This however is not realistic. First, entry into these communities would not be 'free.' It would still need to depend to a considerable degree on social acceptance and personal contact. Second, even if entry were relatively free, it would be difficult for a pre-literate society with a high turnover in membership to define itself by boundaries which have to be demarcated and remembered year after year. How could such territorial claims persist and how could they be transmitted in a pre-literate society with high turnover?[18]

Thus it seems reasonable to say that when a primitive society employs territoriality it defines it socially. This applies not only for the territory of the entire group but also for sub-territorial units within the group's domain because these societies do not possess formal hierarchies and they do not have members who exclude others from resources. This makes sub-territorial partitioning, such as the assignment to families of hunting grounds and garden plots and even of buildings, serve purposes other than that of establishing and perpetuating inequalities. Moreover possessing these types of sub-territories does not place someone else at a disadvantage. The rights of possession are symmetrical. They are used primarily as devices to coordinate effort and to keep people and things out of each other's way. If I am not permitted free entry into your hut or field, I nevertheless am allowed to have such a hut and field of my own and I can deny you free entry into mine.

Figure 3.2 draws attention to several of the relationships between household, community, and resource/land in primitive egalitarian societies and

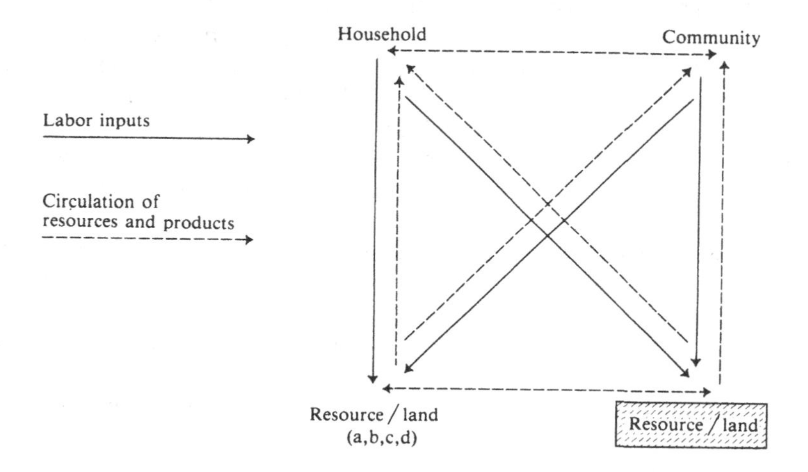

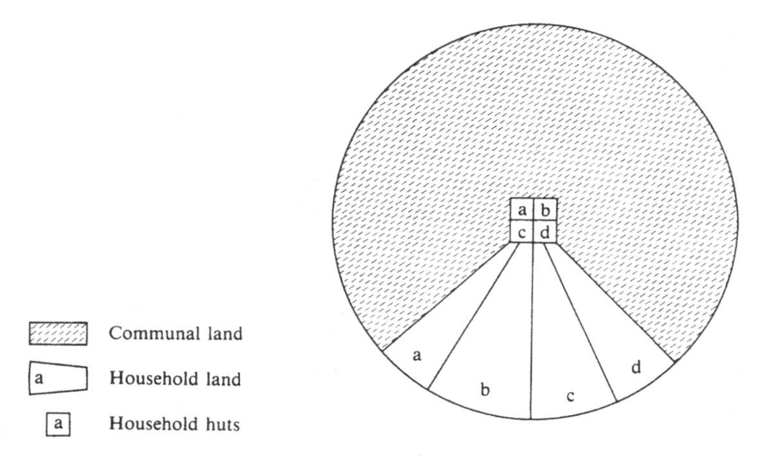

Figure 3.2 Equilibrium in primitive economies: reciprocity among household, community, and resource/land

applies both to agricultural and non-agricultural communities. Each relationship alone or in combination can describe conditions which predominate in a particular primitive society. The solid arrows indicate flows of labor input and the dashed lines indicate flows of output. Land can be either agricultural or non-agricultural. By community is meant any social unit larger than the household, and there may be several such units. Since the system is egalitarian, the flows are reciprocal.

The most characteristic and complete connection, which is indicated in the diagram, is that a household works the land allocated to it by the community. The household is simply holding the land; it cannot alienate it. If there is a change in household needs, due perhaps to births or deaths, then the community will reapportion the land among the other households, or do so less formally by not objecting if free land is taken up or given back by the household. The household works this land (indicated by the solid arrow), consumes what it needs of the produce (indicated by the dashed arrow from resource/land to household) and reciprocally circulates the rest (indicated by the other dashed arrow).

Community land can be appropriated by the household in two ways. Each household may be allowed to work land as it would its own – perhaps by hunting its forests and grazing its animals in the communal pasturage. In this way the household receives directly the produce and consumes some of it and recirculates the rest. Or the land can be communally owned and worked (as in community hunting parties), the product then being consumed by households. Although household land would normally be worked by the household, on occasion, it can be worked jointly as when the community forms work teams to help households clear and weed a field or when a community's tools, such as a large complex fishing net, is borrowed by a single household.

Communities need not use all of these possibilities fully. Some communities may distribute most or all of the land to the households (which still do not 'own' it) and some may have only communal land. And, of course, a primitive community may not be territorial at all, either within the society or between societies, in which case none of the above would apply. Figure 3.2, then, is intended to represent the overall relationships that can obtain between territory and reciprocity in primitive society.

The primitives' use of territoriality supports their basic social organization. When territory occurs at the level of the society as a whole, it is used to prevent non-community members from having access to community resources. When it is used within the community, its purpose is to facilitate reciprocity by assigning different but symmetrical tasks to individuals and households. Territoriality is not used as an abstract mold classifying and separating people and place, but rather as a device to promote their union. Indeed there is little evidence that primitive people see territory, as modern people often do, as a mold or container with clear and precise boundaries that can be conceptually and actually emptied and filled. For them, territory is a place on the earth inextricably tied to events, and the events are intimately and naturally associated with the place.

Space, place, and time

To think of territory as emptiable and fillable is easier when a society possesses writing and especially a metrical geometry to represent space independently of events. An emptiable and fillable space is also a useful concept for a dynamic society. Neither metrical geometries nor repeated social change are part of the primitive world. Society and environment are so intimately interrelated that there would be very little value for the primitive to indulge in speculation about his society being elsewhere or about it having very different spatial and social configurations. Yet such intellectual exercises are precisely what 'modern' social planning and theory require. The nuts and bolts of modern planning involve mental experiments such as 'what if the social order were altered so that land were held differently; what if the village were redesigned, placing this here rather than there, making that rectangular rather than circular, so that certain goals will be more easily attained?' The basis of planning relies on abstracting things from space, that is, on conceptually separating events and space. It relies on our coming to think of space as a system which can exist apart from events and yet be a container for them. The coordinate system of the modern map is ideally suited for the public representation of this sense of space. The grids on the map lay out the surface over which events move. A conception like this becomes sustained not because of its intellectual sophistication but because a particular type of society finds it a useful or significant conception for social action. Modern society is one of change, and geographically this has meant rearranging things in space.[19]

But the political economy of primitive society does not require rapid and continuous changes and hence the primitives have less need for such spatial abstractions. Rather people and place are expected to remain linked for long periods of time and the spatial configuration of settlement and land use will follow customary, traditional patterns. Primitive attitudes towards nature and place reinforce this union. The society derives meaning from place and the place is defined in terms of social and physical relationships.

The same applies to the primitives' sense of space and place in ordinary life. Places are defined by the occurrences of activities rather than by a set of locations in an abstract space. Place in general is felt as involvement with events. Often these events re-occur.[20] The sun, moon, and stars repeat their movements through the heavens and migrating animals return each season by the same trails. Edible plants, berries, and nuts ripen in the same glades each season. These become part of man's encounters with places. Places and events in ordinary, everyday life become linked and infused with meaning; and just as the group sees their society and habitat closely connected, so too does the individual see as intimately interconnected his involvement with events and their spatial configurations. Hence physical space and its properties are not abstracted far from the experiences it contains. A place is an encounter with events.

Without the aid of modern technology carrying people and information great distances with regularity, movement through space would become an adventure with nature providing the traveler with unexpected experiences as he passes from place to place. In this respect, in primitive societies, distances too are experienced as sequences of encounters with events as one moves from place to place rather than as lengths of separation between points. Even when a metrical distance is available, either in units of length, say feet or miles, or in units of time, say hours or days, primitive society would find difficulty in assigning a unit to a particular distance. Such a unit has little meaning in the experiences of overcoming separation. A distance of 50 miles may contain an enormous bundle of varied and unpredicted experiences, some pleasant, some not. These moreover may change each time the journey is made. A 50-mile trip in another direction will certainly contain different bundles of experiences. Or a journey of five days may apply to a particular journey in one season, but not in another. Abstract measurement of distance, then, would not coincide sufficiently with the variety of time, energy, or experience of travel to be of overwhelming value.

In societies with simple technologies, the link of space and time with experience lessens the usefulness of a conception of abstract space, distance, and place. It does not mean that it is not possible to consider them abstractly or to express behavior in terms of such a system. It means that it is not a dominant, socially useful form of conceiving of them. The same is true of time. Again, this does not mean that time goes unmeasured. Rather, as with space, it means that time and experience are so intimately interrelated that their measurements would be reckoned by events that are themselves important and connected with experience.[21]

The trajectory of the sun through the sky, the rhythms of the moon, the changes in the seasons are all natural time pieces which can be and have been readily counted. But it is important to count them only when they correspond with the duration of important events. Thus in hunting and gathering societies it is important to time the seasons because it marks the progression of game, of edible plants, and hence the movement of the people themselves. For an agricultural community it marks the times for tilling, sowing, weeding, and reaping. The cycle of the sun helps to routinize daily activities associated with the care of plants and domesticated animals.

The earliest calendars may have been kept to help mark the time of a human birth. Marshack has argued that the neolithic Venuses have lines around them corresponding to days and phases of the moon.[22] These may have been used to record the stages of a woman's pregnancy. Such a record could tell a wandering band when to stop and prepare for a birth. But it is rarer to find pre-literate peoples keeping accurate accounts of their ages, for no one knows how long one will live, and there are virtually no cases of pre-literate calendric systems that would be used as a frame of reference in which to place long past events. Indeed time becomes hazy once one goes back past the present generation. Up to a point the succession of ancestors may be remembered – but not their duration; then the distant past becomes obliterated or mythologized. Indeed a mythical conception of a distant past is a common characteristic in their world views.

We have included the mythical–magical experience in the context of the ordinary view of space and time because the mythical–magical view of landscape is in large measure an intensification of the ordinary link between space and event. The relationship between the two can be thought of in terms of a continuum measuring the intensity of association between space and time on the one hand and the unique experiences they seem to contain on the other. The more intense the interconnections, the more likely one will become conflated with the other and thus become a part of the mythical–magical experience.

This part of our discussion has been intended to characterize the ordinary (and sometimes mythical) view of space and time in primitive society. It is a view that is held by the primitive in his day to day experiences. But, because these communities are small and static, and their members live close to nature and repeatedly share similar experiences, it is, as well, a characterization of their ordinary public or group meaning of space, place, and time. That is, the contexts of their private encounters become shared and publicly accepted. What is more, this characterization, with some modifications, can apply to the ordinary, personal, and even public experiences of space, place, and time in most pre-modern civilizations. Much of the population in these societies are peasants, anchored to place. Peasants live in small and relatively closed communities, and so here too the private experiences of landscapes and time merge with the public or communal.

This description of the ordinary sense of space and place does not apply without modification to modern experiences. Modern society is more dynamic and involves masses of people from different backgrounds. Technology has made distant places accessible and travel through space reliable, and science has given the public a description of space and time that is abstract and metrical; a description that is useful in controlling nature and that fits more and more the processes of social life. The dynamical quality of society makes it less likely that we can all share the same complex personal experiences. Moreover, the leveling effect of mass culture has in many respects diminished landscape variety and even many of the personal experiences of localities can without much loss be subsumed within the abstract metrical meaning of space and time. Within mass transportation, travel is often experienced privately as duration in time, and places, like suburbia, are so uniform that each can be experienced as location in space. Nevertheless, we still retain private experiences that are far too varied to be reduced to abstract metrical relationships and we often yearn to have even more. But this does not alter the fact that modern society relies on an abstract metrical framework for its public meaning of place and space, and people often find this system consonant with many of their personal experiences.[23]

Civilizations

Even though the models for primitive societies (Figures 3.1 and 3.2) emphasize social stability, these societies change, and a few have been transformed into the earliest civilizations. The word civilized need not connote a change for the better. Civilizations differ from primitive societies in several critical ways. Generally speaking, civilizations are not egalitarian. They contain elites or 'classes' of one sort or another who can extract and consume surpluses and their economies are redistributive rather than reciprocal.[24] Unlike primitive societies, all civilizations are territorial and use territory to help define themselves and their parts. How the first civilizations actually arose, and to what degree they arose independently, is unclear, but major theories point to fundamental processes that were likely to have been involved and which have a bearing on territoriality. These theories differ according to which parts of the model of primitive equilibrium (Figure 3.1) they think were the first to be altered.[25] The fundamental problem they all face is explaining the rise and perpetuation of social economic classes.

Almost all theories of transition assume that the surplus to support an elite was extracted from an agricultural base. Some point to the possibility that the power to extract surplus became concentrated in the hands of an elite because of the need to coordinate efforts to feed an increasing population with a limited resource base. The scarcity in resources could have occurred because of factors such as desiccation or population pressure or both. In terms of Figure 3.1 such an occurrence would alter the low density component. A parallel case is argued by those who see both increased social size and the rise of elites as a consequence of the increasing economic specialization and interdependence of agricultural communities. In this case elites would perform the crucial role of extracting and redistributing resources.

Other theories point to the conquest of agricultural groups by nomads as the primary force in developing elites. This alters the egalitarian/reciprocity relationship in Figure 3.1. Still others emphasize that elites arose by monopolizing some form of knowledge or technology. This would alter the egalitarian component by making 'technology' less accessible. In a primitive society spiritual or religious knowledge indeed could have been an important technology to monopolize. The spiritual pervaded primitive society and spiritual leaders could gain their status by monopolizing powers that were thought by everyone to be essential to the success of the community.

Each theory alone is inadequate as a general explanation and indeed each emphasizes some factors that others place in the background. Our intent is not to argue the relative merits of each or to offer new ones, but rather to draw upon the assumptions in one or two of the theories in order to illustrate the likely changes that occurred in the use of territoriality as the society shifted from a primitive economy of reciprocity to a class-structured economy of redistribution. Although the conquest of an agricultural community by a non-agricultural one may be the simplest model of transition, for our purposes it is inadequate for it does not make clear how the conquerors remain in power and become a legitimate elite. An alternative explanation that can help account for the legitimacy of an elite, and that allows for a range of territorial functions as well, assumes that elites arose indigenously and combined a religious with a redistributive role. Let us see how this alternative can address territoriality.

We know that myth, magic, and ritual were undeniably important in primitive societies and that the earliest civilizations were often led by priest-kings. They collected, stored, and redistributed surplus. Expanded roles of spiritual leaders, coupled with increased economic interdependence among communities, could then well be the key to the rise of a permanent elite which was in charge of surplus and which exercised territorial control over a peasant population. To illustrate the possible steps in such a shift and the changes in territoriality which accompany it, consider the following hypothetical case.[26] Suppose, following along the lines of Elman Service, that a primitive agricultural community 'A' settled in a well-watered fertile plain, as in 'a' in Figure 3.3, situated between a shallow sea and a range of mountains. The land is held communally but partitioned by households for use as in Figure 3.2. A household consumes much of the food it produces, but some is given to kin outside the household and some is distributed to the community through rituals and feasts. Suppose that agriculture is successful and that the community thrives. After several generations, as population increases, residents of village 'A' find new villages, 'B', and 'C', etc., modeled on the same principle as 'A', and these thrive, too.

Each community, though autonomous, may find they could produce some things more effectively than others, so some degree of trade and specialization begins. Importantly, the members of the communities attribute their good fortune to the gods and see their religious leaders as indispensable

a **ORDER OF SETTLING AMONG PRIMITIVE COMMUNITIES**

(A first, B second, etc.)

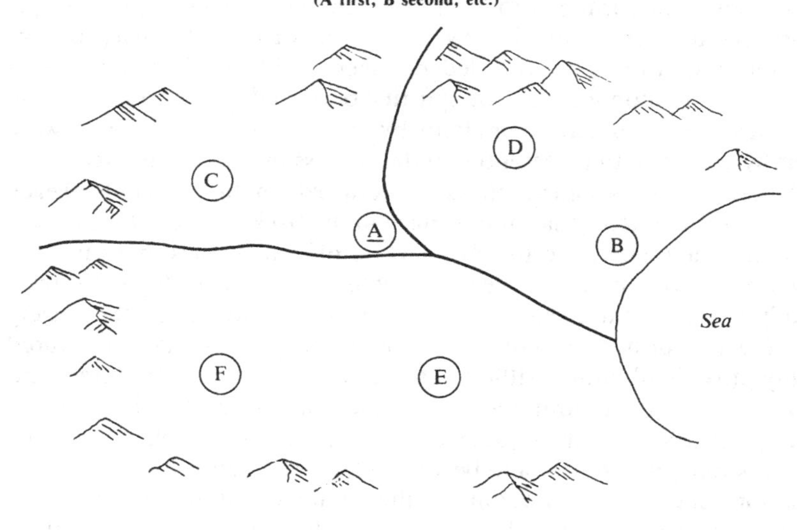

b **INCIPIENT PRIESTLY HIERARCHY AND REDISTRIBUTION AMONG PRIMITIVE VILLAGES**

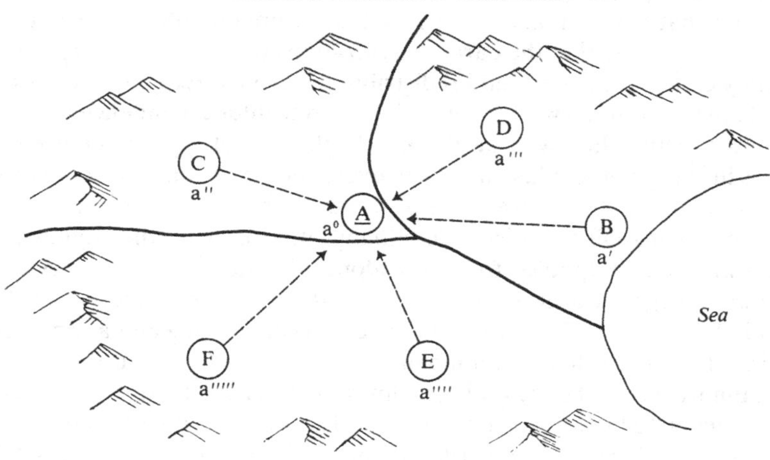

a^0 is the founder of the priestly clan at A, and a', a'' etc., represent the order of male descendants of that clan who became religious heads of their respective villages.

←--- Represents the flow of surplus for religious and ceremonial functions conducted by priests at A

Figure 3.3 Hypothetical order of settling among primitive communities

mediums in soliciting divine assistance. Although leadership in the religious arena, as in other realms, was not originally an inherited position but rather went to the gifted, a crucial difference is that, within a simple society with an accessible technology, religious knowledge and especially ritual incantations and spells offer greater potential to be kept secret than is the case with other forms of specialized knowledge. What is more, key physical and mental attributes of the most effective religious leaders may in fact have been inherited. It is not uncommon for shamans in primitive society to have fits, to hear voices, to have marked physical features. Propensities like these could be passed from one generation to another, and the community may have thought these characteristics to be the keys to spiritual success. Such physical and mental characteristics, coupled with secret knowledge, may have provided the possibility of a powerful means of keeping religious leadership within a family, perhaps from father to first-born son.[27]

Since village 'A' is the oldest, let us assume it is also the largest and most prosperous and that its priestly family may well be thought of as the most powerful. The other villages may encourage their priests to marry the daughters of this family, or perhaps the village asks for the second next eldest son to settle among them to become their local priest. As time passes, the villages become more economically interdependent and the role of religion becomes more and more the responsibility of an extended family or lineage, centered on priests of village 'A' with family branches in each of the other villages (as 'b' of Figure 3.3). The rituals of all the villages more and more become the responsibility of the priestly family.

An extremely important component of these rituals has always been the collection and consumption of community surplus. With the consolidation of priestly power comes a consolidation and centralization of this economic function. The major step in making this spiritual monopoly an economic one is having the priestly family monopolize the role of coordinating and redistributing surplus. It thus becomes the obligation of the priestly family with its headquarters in village 'A' to organize and coordinate the storage and ritual distribution and consumption of surplus for ceremonies. The storage of the surplus as well as the ceremonies for its redistribution are likely to occur at the site of the residence of the head priests of village 'A.' Indeed, the major rituals for all the villages may now be conducted at 'A' (as in 'b' of Figure 3.3) so that representatives from other villages, led by their priests, may converge on village 'A' for major occasions.

As this process increases in scope and centralization, village 'A' builds larger ceremonial accommodations and larger permanent store houses. These are the foundations for the monumental ceremonial centers of ancient civilizations. By this time village 'A' becomes a 'proto-city.' As religious centralization accelerates, so too does economic centralization, and the religious class becomes more intimately involved in the mechanisms of extracting, storing, and redistributing surplus. By now religion and economic power are firmly in the hands of a particular family and its retinue or class. Most important, this concentration can occur with the compliance, if not the encouragement, of the general community. The populace attributes their own success to the priests. If priestly power and glory can be enhanced, so too would that of the common man. This attitude makes many of the members of the community gladly give some of their surplus to this central authority. To do so is to their own benefit as well as to the good of the community. Thereby, concentration of power can occur without physical coercion. If some are reluctant to give, the community may still be small enough for others to notice and for the reluctant ones to be ostracized for placing the rest in jeopardy. Or, if that is not possible, the religious leaders could publicly humiliate the recalcitrants.

As more surplus comes into the hands of the priests, as larger store houses are built, there comes to be the expectation that the granaries be filled, that quotas be met. In order to fill the ritual granaries, at some point the priestly class has to know how much would and could be given by each unit of production, whether they be households or communities. Estimating and organizing contributions requires records, supervisors, and administrators. In other words, it requires the apparatus of a state with administrators and tax collectors. Still there may remain the underlying compliance of the population.

At this point we have a major ceremonial center in a 'proto-city' amidst an agricultural community which has increased in numbers and economic productivity and which now has the equivalent of an economic class inheriting political power and in charge of accumulating and redistributing surplus. Ultimate authority over the land no longer resides in the community or the gods but in the representatives of both – the government and the priest-king. The people may still give over a surplus without overt coercion and life within the villages may remain much as it did before, with the important exception that the villager is now in the position of producing a surplus that will not be consumed directly by the local community. Even if the surplus he produces is no more than before, its extraction and consumption are different. It is more redistributive than reciprocal. But it is probably only a matter of time before the surplus extracted becomes greater than before. Most importantly for us, the extraction of surplus by a central authority, and its eventual heavy burden, means that there will be a change in conception of the land and its people. There will be a greater territorial definition of social relations. The central authority will expect quotas to be met and the peasant's relationship to the land is going to be, in part, imposed. This imposition can take several historical forms, as illustrated in Figure 3.4.

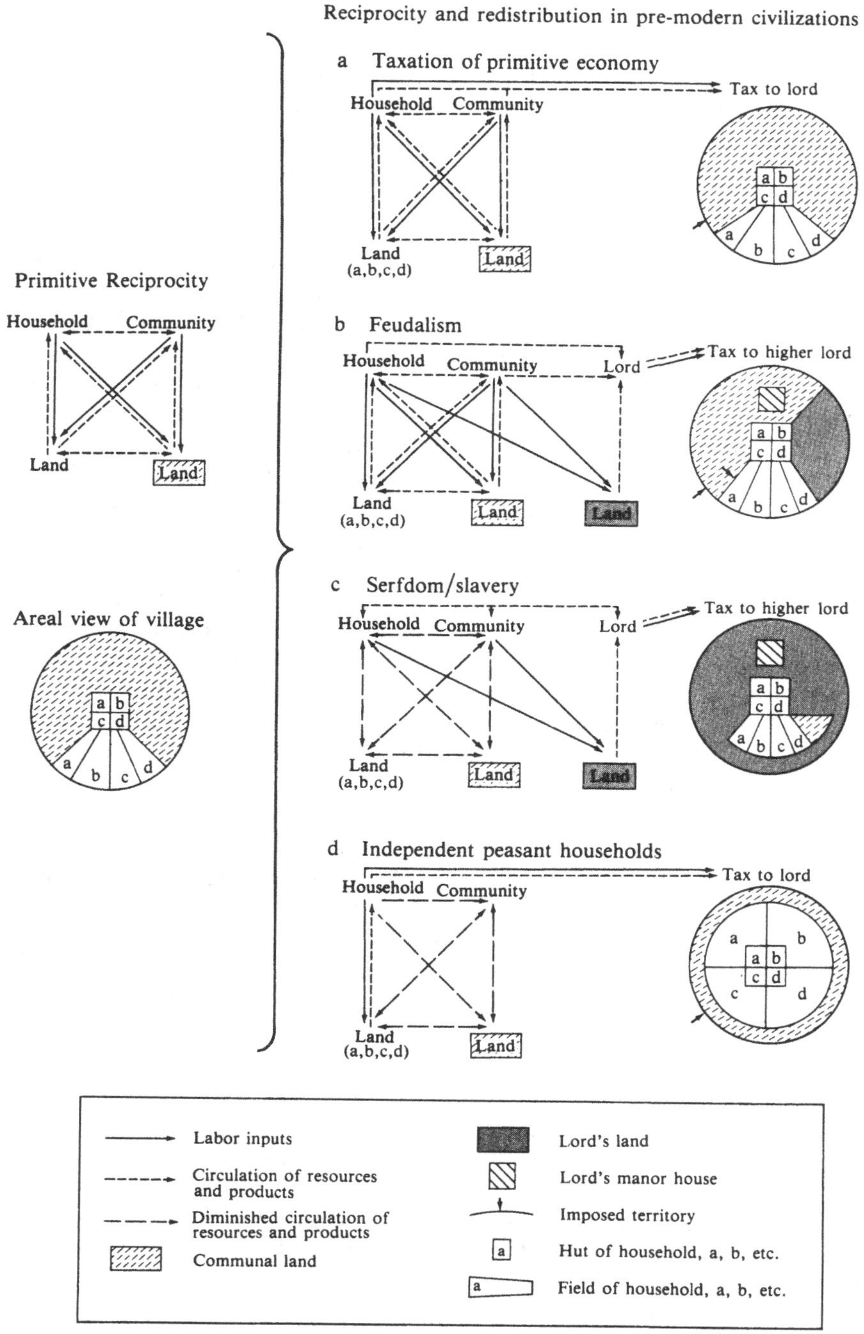

Reciprocity and redistribution in pre-modern civilizations

Primitive Reciprocity

Household Community

Land Land

Areal view of village

a b c d

a b c d

a Taxation of primitive economy

Tax to lord

Household Community

Land (a,b,c,d) Land

b Feudalism

Tax to higher lord

Household Community Lord

Land (a,b,c,d) Land Land

c Serfdom/slavery

Tax to higher lord

Household Community Lord

Land (a,b,c,d) Land Land

d Independent peasant households

Tax to lord

Household Community

Land (a,b,c,d) Land

Labor inputs

Circulation of resources and products

Diminished circulation of resources and products

Communal land

Lord's land

Lord's manor house

Imposed territory

Hut of household, a, b, etc.

Field of household, a, b, etc.

Figure 3.4 Household economies from primitive to pre-modern civilizations

Territorial models

One form, illustrated by 'a' in Figure 3.4, is characterized by the possibility that communal household economies may still be the norm but that the community itself must fill the quota imposed by the central authority. If there are many villages, then these may be grouped within districts. The key point is that in the eyes of the government either the village or the districts become a territorially defined social unit and this territory molds other administrative functions. Perhaps several of these villages become administered by a magistrate appointed by the central government, or the same group of villages must now raise laborers or soldiers. Perhaps the imperial realm becomes governed as a collection of such districts and the land itself, while theoretically the priest-king's, is still in the hands of the village and its peasants. In these sets of options the authority of the state is not intruding far into the life of the community. It is rather overlying a series of primitive societies. This form approximates some areas in Asiatic civilizations which extracted surplus from villages but left their internal organizations virtually untouched.

Another set of options, illustrated by 'b' in Figure 3.4, arises when we consider that a tax on villages like the ones described may well alter their internal relations and destroy their equilibria. Possibly some of the village land becomes allocated to the local priestly clan or their relations. This land may have been held by the ancestors of the local priestly family when land was distributed to all members of the community. Or the priestly land may have been carved from land held by the village in common. Indeed it might appear that the priest-king and his family, as embodiments of the society as a whole, have rights to communal land, and that communal land everywhere becomes part of the priestly demesne.

This land, however obtained, may now be worked by the villages as a way of paying tribute and taxes to the ruling class. This local priestly land may not even be under the direct supervision of the priestly family but rather under the direction of their agents or a local headman. If we consider that the elite may consist not only of a priestly class but also of a secular nobility, and that either can be lodged on a portion of the land they hold (even if they hold it indirectly in the name of the priest-king or some other authority), then we have a model of a state whose authority and representatives are more decentralized and intrude farther into the lives of the peasants. These elements of land holding and authority correspond closely to what has been termed feudal.

A more extreme form of this intrusion occurs when even more of the village lands are appropriated by the ruling class and the peasants are simply serfs on vast estates as in Eastern Europe during the sixteenth to the eighteenth centuries (see 'c' in Figure 3.4). The most extreme form – slavery

– occurs when the workers have no liberties at all, when they are owned and when neither land nor tools are left in their possession. Most of the work day must be organized by the slave holders and they must provide the slave with the necessities of life.

Yet another possibility ('d' in Figure 3.4) is that local community connections may weaken and the peasant households come to 'own' their own land, leaving society with relatively independent land holders – some of whom may become richer than others. A weak central authority may still attempt to extract taxes but the society is composed primarily of independent farmers, the richer of whom may employ or enfief the poorer, or even hold slaves. A situation not unlike this occurred in Viking society.[28]

Whereas one of these types may characterize a particular society at a particular point in time, all could exist to varying degrees together.[29] Moreover, in all but the slave system, the peasants individually or communally 'owned' most of their own tools and implements. They were left to support themselves, and in so doing were in control of much of the work process. Even when they worked for the lord of the manor or the estate, they were often left in charge of the details of the labor process and used their own implements to fulfill their labor obligations. Surplus was extracted, and thus people worked also to support an elite but the rhythm of work and its details were mostly in the hands of the peasants.

We have concentrated on the relationship between lord, land, and peasant because the wealth of pre-modern societies was based on agriculture. But it is important in Figure 3.4 to take notice of the position of the merchants, who were present to varying degrees in all civilizations. Their lives, like those of the peasants, were regulated by the ruling class. In the case of type 'a', most of the merchants would have been concentrated in the capital cities, providing scarce and expensive commodities for the ruling classes. Not only would their customers be the wealthy, but the political elite or government very likely would have licensed and taxed these merchants and controlled the prices of their commodities. The concentration of wealth in the imperial cities, moreover, also would have drawn to them the artisans and craftsmen.

In the case of type 'b' – feudalism – the merchants would have catered to the nobility but because they were dispersed, and often living among the peasants, there may have been a greater decentralization of the control of trade. Even the feudal peasants may have had greater access to some of the less expensive goods. Merchants would have also visited the nobility on the vast estates in serfdom or 'c.' But it is not likely that the merchant would have had much direct contact with the independent peasant households in the types illustrated in 'd.' Rather they would have more likely remained in the cities in which the elites resided and established occasional fairs and markets in the rural areas.

Shifts in space, time, and territoriality

Knowing which one of these relationships predominates in a particular case helps explain the distribution of land and power and some of the details of spatial organization. Yet all of these types share overall territorial effects. The sense of a community holding land in trust from one generation to another is replaced to varying degrees by the notion that the ruling classes (however else they may differ) are embodiments of the people and the gods and are also the trustees of the land; and land, as the basic source of wealth, is a territory to administer and a place to which people are bound. Binding the villager or householder to the land makes him a peasant or serf.

From the householder's point of view, the land he works and the village in which he lives may still seem to him as a natural entity, as the place to which he belongs emotionally and spiritually. He may not be terribly conscious of the fact that his socially defined territory is also an imposed territory – forming a territorial definition of social relations. Even if the boundaries of his territories are in some sense artificially created or imposed from above, the imposition of territorial authority may not have affected his personal sense of belonging to a place. Unless the creation and content of territories change extremely rapidly and repeatedly, there is a tendency of a people to try to make a place feel like home. This means seeing it as a socially defined, natural territory. It may, however, require considerable effort to gain such an attitude and even then it can be more fragile and less enveloping than the social definition of territory attained in a primitive society that has no authority above it.

Whether the peasant views himself as part of an imposed territory or not, the crucial issue is that, from above, the central authorities do. To them the peasant is the embodiment of the instrument that makes land productive. Land would be useless without the peasant. The peasant becomes bound to the land. In this regard, as well as in other ways, the central authority partitions its empire and people territorially, classifying and controlling groups, both large and small, through the control of area.[30]

Thus the rise of civilization makes formerly internally defined territories also serve as molds for other forms of social relationships. A group of villages (whether, according to Figure 3.4, of types 'a,' 'b,' 'c,' 'd,' or combinations thereof) becomes an administrative area established for raising revenues and these areas, as in Figure 3.5, become administrative units for relief in cases of plagues and famines, or for the provision of soldiers in time of war. Territorial boundaries could be used deliberately to divide and subdue or conquer hostile communities by cutting through older 'natural' community areas. In all of these cases, territoriality is being used hierarchically and asymmetrically. Control of territory is in the hands of a ruling class or their representatives. The greater the number of territorial

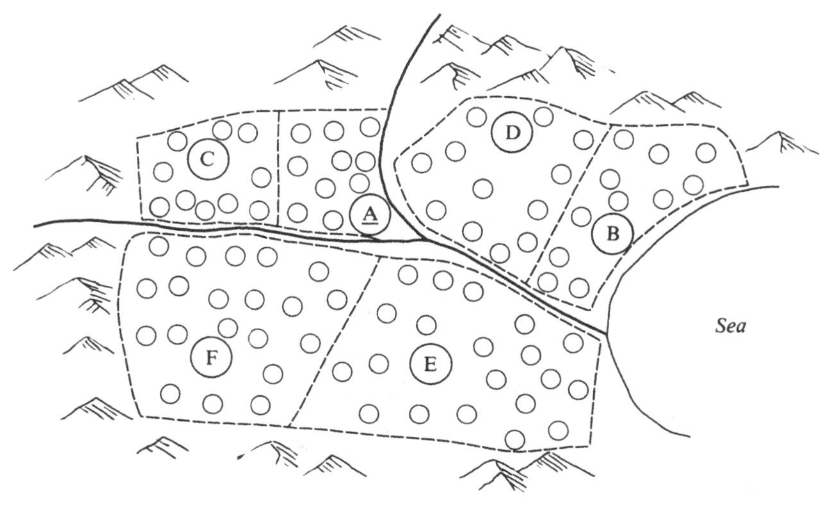

○ Village territorial units socially defined and recognized by central authority at A

---- Territorial administrative boundaries imposed by central authority at A and administered by regional centers B, C, etc.. These administrative units define groups of villages territorially.

Figure 3.5 Development of territorial definitions of social relations and territorial hierarchies

hierarchies, the more likely it is that the boundaries are artificially imposed. And the more the administrators are rotated from territory to territory, the more territoriality produces distant and impersonal relationships between governors and governed.

In pre-modern society the land was the primary source of wealth, and agriculture the principal occupation. But it should not be forgotten that the rise of civilization also occasioned the rise of the city and complex architecture. Here, too, are found multiple territories that mold events and hierarchically and asymmetrically define social relations. Temples and palaces for instance are most often internally subdivided to mark off degrees of the sacred. The more sacred the place, the less accessible it is to the average person. Only the highest priest was allowed to enter the holy of holies and only the royal family and their retainers were allowed within the heart of the palace (Figure 3.6).

Perhaps the most striking example in pre-modern civilizations of a territorial definition of social relationships is the use of territory to define the entire political community, state, or empire. This use of territory, though important for empires, was by no means precise by modern standards.

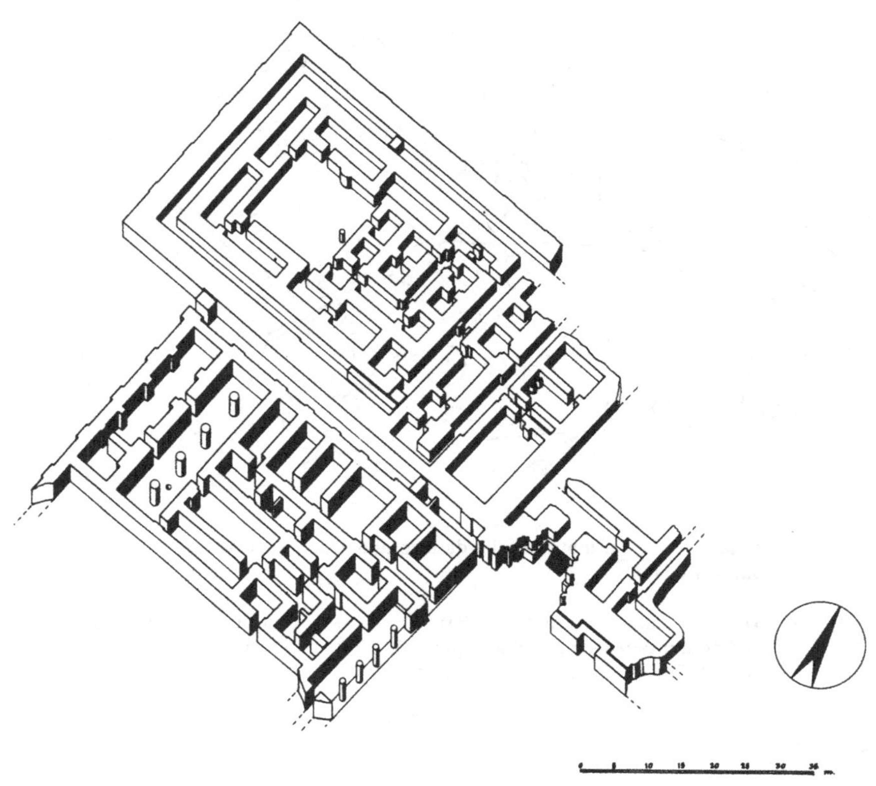

Figure 3.6 Territorial subdivisions within buildings: early dynastic palace at Kish
Source: E. L. Mallowan, *Early Mesopotamia and Iran* (London, 1965). Reproduced by permission of Thames and Hudson Ltd.

Boundaries were never delimited as accurately as they are now because surveying and mapping techniques were limited, and records of land holdings were not kept uniformly and accurately.[31] A lord may have known that a village and its land belonged to him, but he and most of the villagers may not have known exactly where village land began and ended. Yet a general sense of extension in space was clearly part of the definition of every civilization.[32] Emperors controlled a people within a territory. Indeed the emperor controlled anyone within the territory. A people were socially diverse, yet territorially defined.

These territorial claims, as well as other details of imperial possession, were often represented by maps, sometimes drawn to scales and with specific projections. Time, too, was often presented metrically by complex calendric systems. These maps and calendars were certainly more elaborate and abstract than those found in primitive societies (primitives have drawn maps, but not metrical and projective ones), and in this regard they may indicate a qualitatively different level of spatial and temporal awareness, at least on the part of the elites who employed such devices, than existed before the rise of civilizations. This change in the representation of space and time between primitive societies and civilization was not, however, as profound as was the change in the use of territoriality. (And, as we will see later, a far more fundamental shift in the meaning of space and time was to occur with yet another fundamental change in territoriality at the time of the Renaissance.) The peasant in pre-modern civilizations conceived of ordinary space and time in much the same way as did the primitive. And even the more abstract calendars and maps of imperial officials were ladened with mythical–ritualistic meanings. Calendric systems of empires had astrological overtones. Priests were the reckoners of time, and time was reckoned to keep the empire in harmony with the heavens. Symbols of geographic space played an especially important role in establishing imperial harmony. Space was given mythical content. For example the division of the Chinese Empire into four quarters was conceived of as a mirror of cosmic order. Even the apparently modern metrical coordinate system found in third-century Chinese cartography was infused with mythical meaning. Each square in the grid was a microcosm of the whole world, and each side of the square contained part of the meaning of the direction it faced.[33]

These mythically infused conceptions of space and time were also used extensively in rituals of government. To rule such vast empires containing different types of peoples meant relying heavily on ritual. To ensure their success, new cities were often located and designed according to mythical space and time. Investitures of rulers required ritual connections between capital and empire. Agricultural prosperity could require that the land be ritually fertilized and ploughed by the gods, in the symbolic form of the emperor.

Often ritual space and time were interrelated. Even the conceptions of space and time in the Classical World were infused with content. Greek geometry was about two- or three-dimensional objects and not about abstract, empty space.[34] Greek philosophy of space was not about an empty metrical continuum.[35] The Greeks did not for instance develop projective geometry.[36] Ptolemy's Cosmographia came close to an abstract representation of space, but his projections were about solid objects and his geography was closely linked to astrology. Finally, these representations of space and time did not constitute the day to day experiences of the peasant. For him, ordinary space and time were still limited to local and personal experiences.

Capitalism

We have glossed over important differences among pre-modern civilizations in order to focus on the major changes in territoriality. But we should add that, not only were pre-modern societies different, only one of them gave rise to capitalism and the modern state. This must mean there existed crucial differences between Western Europe, where capitalism began, and the rest of the world. But what exactly they were remains an open question.[37] Far clearer are the characteristics of the actual economic system and changes in conceptions of space, time, and territoriality that first merchant capitalism and then industrial capitalism initiated. Again it should be noted that whereas capitalism and modernity are not synonymous, the former was one important component of the latter.

It will be recalled from Chapter 2 that important among the uses of territoriality which can be expected primarily in capitalism and the modern era are the sense of an emptiable space, the increased uses of territorial hierarchies to further impersonal relationships, and the use of territoriality to obscure sources of power. Here we will consider the interconnections of these effects within capitalism, concentrating primarily on the first, because it is both historically prior and necessary for the others. The point we will make is that the repeated and conscious use of territory as an instrument to define, contain, and mold a fluid people and dynamic events leads to a sense of an abstract emptiable space. It makes community seem to be artificial; it makes the future appear geographically as a dynamic relationship between people and events on the one hand and territorial molds on the other. And it makes space seem to be only contingently related to events.

Mechanisms

It is generally agreed that an important requirement for the rise of capitalism is an extensive market system for buying and selling commodities. But this can hardly be sufficient, for traditional societies contain both merchants and markets and yet, in these pre-modern cases, trade does not fundamentally alter and displace subsistence household economies. Rather merchants are grafted on to them. In addition, then, to a commercial segment, capitalism needs to make labor and capital dependent on commerce. This dependence transforms the role of the merchant from a person supplying conveniences to one supplying necessities, and it transforms the role of the peasant from one relying on himself and his community for subsistence to one relying on the market and the merchant. These are characteristics of what we will term merchant capitalism and are illustrated in 'a' of Figure 3.7. Many Marxists may not consider the merchant capitalism a distinct mode of production but rather a transition to or part of capitalism.

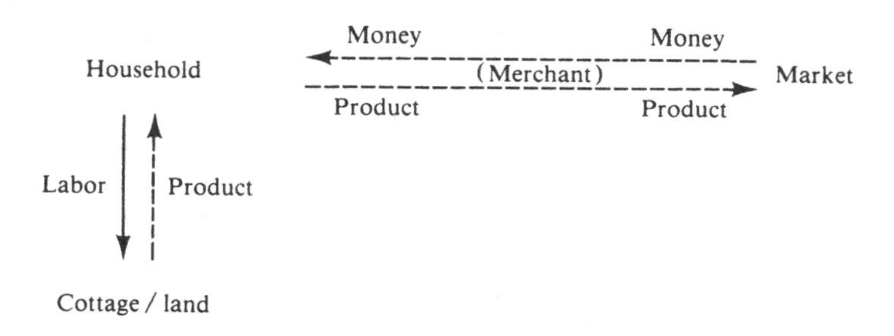

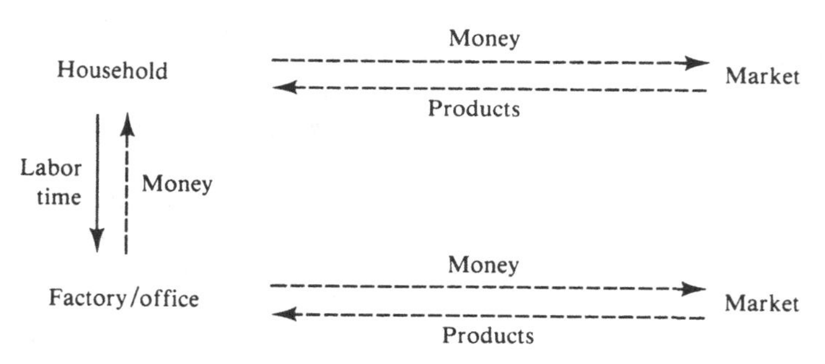

Figure 3.7 Households in merchant and industrial capitalism

One way for merchant capital to take hold is to 'free' peasants from the land so that they can enter into the market while making sure they cannot have the option of returning to subsistence or traditional livelihood if commerce fails.[38] The sale of their labor or their products must become essential to their livelihood. Traditional economic obligations and community relations must be severed, and the land of the typical freed peasant, if he has any, must be insufficient to provide him with the alternative of subsistence. He must engage in market-related work either by leaving the land or by working in or nearby his cottage to supplement his meager produce from his small fields and gardening. If he has the option of ignoring the market by living off his own land or by finding enough support from communal land then his connection to the market would be only voluntary. The market would provide only a commercial trapping to a peasant livelihood.[39]

The alternative of subsistence must also be taken away from the major land holders. This means they need to be prevented, by social convention and by more lucrative alternatives, from working their own lands and thus becoming peasants, and from attracting peasant workers to work their lands within the older customary relationships of feudalism and serfdom. The landlord must be placed in a position such that his 'rational' choice is to purchase 'free' labor, and this means that his land must be producing for the market. An extensive and penetrating commercial network in place, as had portions of Europe in the thirteenth and fourteenth centuries (as in the Hanseatic League), was certainly a critical part of the shift to capitalism. But in itself, it did not 'free' labor or capital. The details of how this transition from feudalism to merchant capitalism was accomplished varied from place to place in Western Europe.[40]

Part of the geographical differences among merchant capitalism, industrial capitalism, and pre-capitalist economies are indicated when 'a' and 'b' of Figure 3.7 are contrasted with Figure 3.4. In merchant capitalism ('a' of Figure 3.7), the individual household remains a unit of production. Work occurs in the household and most often on tools or instruments that the household owns. The key difference between what occurred before in pre-capitalist modes and now in capitalist ones is that the dependence on the market severs the household's connection to a local subsistence community. The household often has to purchase raw materials from the market, bring them home to work on, and then sell the finished product back to the market in order to receive money to purchase the household necessities. The merchant, often serving as the intermediary between household producer and market, makes his living by buying cheap and selling dear. He does not normally own, organize, or supervise the production process (and in this regard Marxists would say he does not extract surplus value). He simply distributes raw materials and finished products. The householder, to some degree at least, remains in control of the internal organization of the work process. He decides how it is set up, and at what pace work occurs.

In industrial capitalism ('b' of Figure 3.7), however, the worker no longer owns his own tools or works in his own home. Now he has to work on the tools and in a work place, such as a factory or office, owned by a capitalist who organizes, supervises, and controls every facet of work. No longer does the worker sell his own product to the market. Now he sells his labor time to the factory owner. The uses to which that time is put are determined by the owner. Work and home often become physically separate and territorially distinct entities. And, with husbands and wives having to work away from home often for long hours, other functions of life become territorially segmented. There arise schools to care for and educate children, and hospitals to care for the sick. These other territorial separations and segmentations accompanying industrial capitalism have had profound effects on how territoriality is used in capitalism and will be addressed later. Here it is important to emphasize that the commercialization of both labor and capital within both merchant and industrial capitalism result in a complex shift in the locus of political force and the way it is perceived. The characterization of this shift depends to a great degree on one's ideology.

In traditional or class divided societies, political power and government were unambiguously in the hands of the elite. Power was highly visible and coercive. Whatever other functions may have been performed by the state, its ultimate purpose was to maintain the elite by enforcing the extraction of surplus. But in capitalism, political power and the state appear in a different light. They seem to be potentially neutral forces. Labor and capital, as factors of production, enter into a wage-labor agreement or contract. Supporters of capitalism interpret this as an arrangement kept free and equitable in the long run by the forces of supply and demand in a competitive market system. If labor thinks it is paid too little, it can bargain for more or go elsewhere, and capital, too, can hire the labor it chooses. In industrial capitalism, capital can move to new locations to find new labor and can substitute more capital and materials.

To the supporters of capitalism the state is or can be a neutral agent, impartially enforcing these and other contracts and agreements. In this sense, the economic and the political appear to be independent of each other. The economic is the private sphere, which, in industrial capitalism, is based particularly on the wage-labor contract and governed by the invisible hand of the market. The political is the accessible public arena for resolving conflicts according to equitable rules and procedures. It assists the economic primarily by ensuring that agreements are upheld and by providing public goods and services such as roads, canals, defense, and justice. In return it receives legitimation from these recipients and support through taxes. Whereas the most ardent supporters of capitalism may argue that the powers of the state are neutral and ought to be limited, they also realize that in fact the role of the state is vast and most analysts would agree that the state affects large segments of life.

This appearance of political neutrality and quasi-independence from the economic is a fundamental difference between the state in capitalism and the state in pre-capitalist or class divided societies. However, the supporters of capitalism and their Marxist critics disagree whether it is more than an appearance. The Marxist argues that, particularly in industrial capitalism, the inherent oppressiveness of the capitalist system is based on, and yet obscured by, the wage-labor agreement. Capital owns the means of production, yet labor creates value. In order for labor to work it must give over to capital some of the value it creates. The labor contract is the device by which capitalism hides the extraction of surplus value under the guise of a free and open agreement. In other words labor is paid less than it is worth. Capital extracts this money simply by having the power to withhold jobs. This inequality, and the extraction of surplus, is disguised by the apparent freedom of labor to enter into a labor contract. Through this contract, surplus is extracted without the use of overt force and without overt political intervention. Yet by providing support for this contract and for private ownership of the means of production, the state, under capitalism, is committed to supporting this unequal relationship.[41]

Not surprisingly, the differences between the Marxist and the non-Marxist views of political economy and the state influence their respective evaluations of territoriality. The major issue is that Marxists would tend to see political territorial units as creating and legitimizing inequalities and obscuring sources of power, whereas non-Marxists would not. We will focus on this issue later, but for now, it is important to recognize that whereas the two views can see territoriality employing different sets of effects, not all of the interpretations of capitalism's uses of territory fall neatly within a Marxist/non-Marxist division, and that the two views can coincide to a considerable extent concerning some of the basic trends in the relationships between the rise of capitalism and the state on the one hand, and conceptions of space, time, and territoriality on the other.

Both would agree that in addition to presenting the state in a more neutral role, capitalism has increased the geographical mobility of the general population. With trade and mobility came a geographic extension of political power. This helped secure access to new markets and raw materials and helped maintain reliable transportation and safe conduct within the domain. (National integration was in the interest of the rising bourgeoisie and also of the monarchy. It provided the monarch with a direct source of revenues and it helped bypass the local feudal interests that diluted the monarch's authority.)

With capitalism came the need for ever increasing accumulation and hence of innovation. Both a merchant and an industrialist must continue to expand their markets, to sell (and produce) more and more new goods. If they do not, they run the risk of losing their businesses to others who will and who can undersell them. Thus change becomes an imperative of the system, and the rationalization or ideological justification for change becomes embodied in the idea of *progress*.[42] Time, it is believed, will lead to change, a change for the better. This abstraction and attribution of value to time becomes more intense with the growth of wage-labor and industrial capitalism. It also becomes increasingly interrelated with the changes in meanings of space and territory.

Ideology

Much of Renaissance thought reflected the needs of the newly emerging economic system. Man took center stage and man's sensations were to be the primary sources of his knowledge of the world. Placing experience, rather than scripture, tradition, and authority, as the font of knowledge opened the world up to an experimental and even scientific outlook.[43] New needs and experiences could be met by travel and also by new commodities. The exhaltation of wealth and consumption (which the new Protestant denominations heralded in their idea of a 'calling') is as important a component to the success of capitalism as are competitive business acumen and access to resources and markets. Helping to make the accumulation of wealth and new products possible was the market mechanism of supply and demand which provides a powerful device for revealing and measuring needs. The market mechanism quantifies worth or value, and as the realm of market activity expanded, more and more things became valued in terms of their market price, rather than by their traditional value or usefulness.

Quantifying value was paralleled by the use of measurement in science. Facts could be described in terms of such quantifiable units as locations, sizes, and weights, just as goods could be described in terms of prices. Experience of all kinds was thus becoming more amenable to measure. Using quantitative terms to describe empirical reality and economic value was far more useful to a dynamic society than was describing them in terms of traditional or customary uses that were no longer shared, or upon authority that was losing its legitimacy. But quantification had yet another far reaching consequence. It helped to make many realms of Western thought more abstract and self-critical.

Representing reality through numbers is not itself intrinsically more abstract than representing reality through words or pictures. But when it is encompassed within a calculus of quantitative relations such as a geometry or an algebra, it then reinforces a form of representation that is precise and conventional, yet cold and remote from the complexities and contradictions of ordinary experience. Such clarity and remoteness are characteristic of scientific conceptions in general. The meanings of scientific concepts are consciously created and arrived at by the consensus of the scientific community. They are not seen as natural symbols, but rather as conscious constructions. This consciousness encourages science and related realms to become self-critical of its own representations.[44] Quantification enhances the clarity, precision, and public nature of scientific terms. Representing nature through well-defined, specific, and often quantifiable terms or symbols facilitates a prolonged and complex conceptual separation and recombination of parts of reality. It makes scientists aware that they are creating symbols whose meanings are negotiable. This conscious conceptual activity adds to the self-critical nature of science. It promotes an awareness that one is building models of reality and not creating reality or affecting the reality represented. Anything one does can affect reality but the scientist recognizes that the symbols used do not themselves manipulate directly the things they stand for (as is believed to be the case in ritual and magic). And this consciousness of the role of symbolization, assisted by the use of quantification, helped increase the spirit of intellectual experimentation in other modes of thought.

The role given by the Renaissance to human observation and experience, the quantification of facts and values, and the emergence of science, were by no means the direct causes and effects of one another. Nor did they immediately dominate Western thought. They were consonant with ideas which were to interpenetrate and reinforce one another as capitalism developed. They were also interrelated with the changes occurring in the conceptions and uses of earth space, territory, and time. Indeed, the development of a more metrical and quantitative conception of space, place, and time had the paradoxical effect of facilitating movement, coordination, and control of activities over the earth while removing the meaning of space and time from much of ordinary experience. Space and time have become abstract frameworks to which events and experiences are contingently related. This abstraction has greatly affected modern Western ideas of reality. It has helped to make much of Western thought self-critical; it has helped to diminish ritual and to increase skepticism.

Space, time

Capitalism is the first truly dynamic economy. The changes that capitalism creates are rationalized through the belief in progress. People expect that the future means change for the better. True, often other societies hold that the future might be better, but this usually means either that there are cycles of good times and bad, and what has been will eventually occur again, or that something entirely new and revolutionary will occur, but only through divine intervention. Belief in the Western notion of progress, however, means belief in a more or less continuous secular change to new, better, and heretofore unavailable and even unforeseen conditions. Blind faith in progress, like blind faith in anything else, is an unsophisticated use of an abstraction. But there were firm grounds in post-Renaissance society for a more cautious belief in progress if that is taken to be as a generalization about the changes that could be expected in material life.[45]

The idea of progress, as well as more mundane experiences of change in modern society, rely on an abstract metrical notion of time. Measuring time's passage still requires measuring changes in the material world, like the path of the sun. However, the primary difference lies in the fact that the measure can have unrestricted referents. It can refer to familiar events such as the movement of the heavenly bodies, or to the ages of people, but its primary function becomes simply to serve as a neutral measure to mark duration and sequence of any and all events. In fact instead of events marking time, metrical units of times have come to define events, as when in factory life work begins at 8.00 a.m. and ends at 6.00 p.m.

Metrical time became a more important part of a world frame of reference with each new advance in mechanical time pieces. These permitted time to be measured in ever smaller, equal, and precise units. These units, as well as the older divisions of days, weeks, and years, began to have value in and of themselves, as more and more work came to be reckoned in terms of buying and selling labor time.[46]

As metrical time was becoming freed from any particular experience or context, so too was geographical space. Few societies, including medieval Europe, made systematic use of a spatial grid to represent the world. The Chinese made use of a coordinate system but it seems to have been infused with spiritual meanings. Greek geometry and cartography were more concerned with solid objects than with abstract space. Yet some Greek mathematics and cartography had the potential to address earth and celestial space itself as a metrical system. Ptolemy's discussion of projections and coordinates comes as close as any pre-modern formulation to an abstraction and metricization of space. In this regard it is significant that Ptolemy's system was available to the Christian Middle Ages by A.D. 1150 but was not used by medieval geographers. Rather, medieval scientists applied it to the heavens and even there it was interlaced with astrology. For earth space, medieval geography often substituted a mystical view derived from Church doctrine. In one important series of medieval representations, illustrated in Figure 3.8, the earth was divided into three regions, Africa, Asia, and Europe, and these were joined often at the center in Jerusalem. The import of this conception was that wherever one actually is located in physical space is immaterial unless one is at the center in the heavenly city, Jerusalem.[47]

Not until the fifteenth century with the awakening of interests in navigation and trade was Ptolemy's system of mapping the earth rediscovered. The year 1405 saw the first translation in the West of Ptolemy's geography. Soon thereafter representation of space in terms of coordinates such as longitude and latitude became the standard cartographic conception of space. Using a coordinate system in and of itself to represent space depends on imagining the globe not as an amorphous topography but as a homogeneous surface ruled by a uniform grid.[48] An abstract metrical space was joined to an abstract metrical time to form a general, yet precise, frame of reference in which human experiences are contingently located.

Mapping space in terms of coordinates was only one of two primary instruments to express the awareness of space as an abstract framework for events. The other was the fifteenth-century discovery of perspective painting. Before this time the concept of space in painting was overwhelmed by position and size. The size of an object and its position in the painting indicated something of its importance but nothing of its actual location in a geographic reality. Within the context of perspective painting the events to be depicted were literally painted into a pre-existing coordinate system that

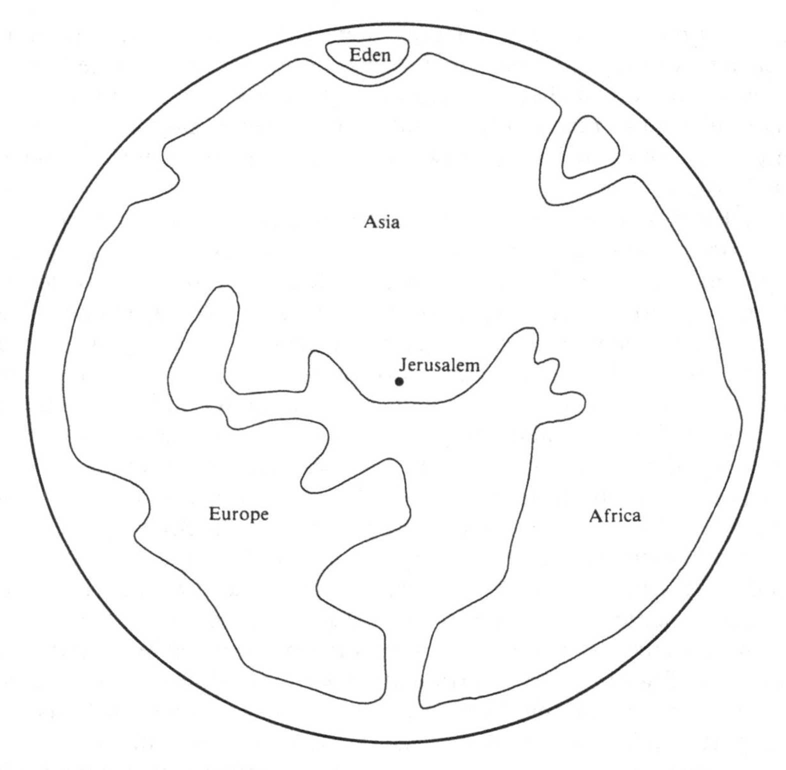

Figure 3.8 Schemata of T-O map

represented space itself. Perspective painting and Renaissance cartography reinforced each other. Artists were aware of new cartographic methods and cartographers were often artists. Their interconnections may have been so close that Ptolemaic rules for map projections in the Almagest may have been adopted by Alberti, one of the founders of perspective painting, in his constructions of perspectives.[49]

Developments in cartography and painting were of great importance in conceptualizing and rendering an abstract spatial system. But employing the system for practical ends encountered several impediments. The Ptolemaic gazetteer which was used by the Renaissance as the basis of the map of the known world was considerably inaccurate. Latitude was fairly easy to measure accurately even in the Ancient World – by observing the height of the Pole Star or the sun – but longitude, which required the simultaneous observations at different geographical locations of celestial events, was not possible to measure with any degree of accuracy until reliable, portable, time pieces were developed in the mid-eighteenth century. Until then, only rough approximations could be made.

Despite these temporary limitations to the applicability of the abstract metrical spatial framework, it was found to be useful enough to allow new developments in ship design and navigation to be put to use in exploration, and these technological innovations in turn helped to improve map accuracy.[50] More and more things became geographically accessible and accurately locatable. But in a complex world, fueled by capitalism, territoriality is required to clear the way for geographic accessibility. And several developments needed to take place before the abstract spatial system could be incorporated into and matched in its abstraction by a modern use of territory.

Territoriality

A modern use of territory is based most of all upon a sufficient political authority or power to match the dynamics of capitalism: to help repeatedly move, mold, and control human spatial organization at vast scales. This modern use of territory is first a matter of degree and intensity. But at some point it begins to lead to a qualitatively different sense of territory and space. Territory becomes conceptually and even actually emptiable and this presents space as both a real and emptiable surface or stage on which events occur.

From the beginning of recorded time, there exist cases of territorial definitions of social relations: of molding people to form communities and of abstract plans for new towns and colonies. Everywhere, examples can be found of the creation of new political territorial units from older ones and on a smaller scale of holding land as unencumbered property. These territorial definitions must have made it appear to some degree as though land was simply an area of space. But these hardly constituted a majority or in many instances a significant proportion of society's uses and experiences of territory. On the contrary, the vast majority of communities were established so very gradually and were of such long duration that even if territoriality was used to initiate a settlement, with the passage of time, the place came to be thought of as a natural and unified entity.

At the political level the Old World neither encountered vast empty tracts of land nor had the need or the ability to create them. Even when one group conquered another, the conquerors, with few exceptions (such as the Mongol effort to clear the Chinese from the steppes), did not attempt to expunge the pre-existing communities and their land holding relationships. For any land to be of value, it had to be occupied by cultivators and most land in the Old World was so occupied. To displace resident populations, even if a monarch or state had the power to do so, meant that refugees would only have to be replaced, and this was a tedious and costly undertaking. Old World conquest before the Age of Exploration involved subduing and establishing suzerainty over older and resident agricultural populations. It did not involve displacing them entirely, even when these populations were thought by their conquerors to be inferior.[51]

Even though medieval Europe, like all civilizations, employed a territorial definition of social relations to some degree, the predominant conception of territoriality in the Old World before the Renaissance was of a social definition, and the transition to an awareness of a territorial definition to accompany the rise of capitalism would have been far more gradual if it were not for the discoveries of the New World. The New World, and especially North America, presented European powers with a vast, distant, unknown, and novel area. This meant that with the limited technology and political power at their disposal, Europeans could still 'clear' much of the space and form territories at all geographical levels, with an intensity that was impossible to match in the Old World.

The discovery of the New World accelerated the use of a territorial definition, but it took time for this sense to match the abstractions of the cartographic representation of space. Moreover, the increasing intensity of a territorial definition along with the use of abstract metrical space did not replace, but rather added to, former conceptions. Indications of the changing attitude toward territory and space are found in many facets of New World political life, especially in North America. Evidence for thinking of territory as emptiable space is indicated early on in North American charters and grants which delimited their claims by using the abstract metrical lines of latitude and by their provision for a hierarchy of administrative sub-territories long before the land was surveyed and settled. It is indicated by the Europeans' claims that the land was virtually uninhabited, and by the displacement of most of its aboriginal population, some to reservations. It is found in the gradual change of definition of community, from one in which new settlers were admitted only by consent of the community, to one in which admission required only residence within the community's territory. It is indicated by the shift, from a form of representation in which a community was thought to be an organic entity with a common interest and thus needed only a single representative to give voice to its needs, to proportional representation based on periodic censuses in which the community is thought of more as a collection of individuals than as a unified body. It is found in the United States Constitution in the use of areal representation as a device to divide factions and balance power, and it is found in the territorial partitioning of the Western Lands.

The cases of treating territory as an emptiable and fillable mold multiply as we approach the present and the same trend is seen when we look at territories in smaller geographical scales within neighborhoods and buildings. The primary difference is that here, unlike the political level, which, because of coordinate maps, could be presented as an empty space to partition and then fill, the small-scale and architectural levels were seen as emptiable only after they were thinned out; only after each thing was put into a separate place. This thinning out did not mean a lowering of density. On the contrary, it meant first isolating and segmenting the specific activities to be contained, and this often meant a multiplication and intensification of territoriality.

Before this transformation, medieval Renaissance town streets (and those of other pre-modern societies), for example, were filled with the hustle and bustle of numerous activities. One could find merchants peddling wares, beggars panhandling, families socializing, town criers spreading news, and public trials and hangings, occurring together in streets and squares. But as commercial interests became more important and as the effects of commercial capital uprooted more and more peasants, there developed greater restrictions on access to public spaces like roads and squares. Rules were enacted prohibiting merchants from hawking their wares on the streets, restricting beggars to certain locations, prohibiting social gatherings, and in general limiting the use of streets and roads only to transporting people and goods from one place to another. As the activities in streets became thinned out and cleared for transportation, the city itself became more and more economically differentiated, water fronts were cleared for warehousing, stock exchanges were established in accessible areas; and, with industrial capitalism, the city took on its modern form of residential areas and manufacturing and central business districts.[52]

A corresponding thinning out of place through isolation and segmentation can be found at the level of domestic architecture. Before the commercial revolution the major architectural organization of even the largest houses was not based on a careful assignment of specific functions to specific rooms or places. Rather, homes were subdivided into rooms that could serve multiple functions, and in most cases each room led into another, for few central corridors or hallways existed solely for access.[53] But just as activities in the streets and shops became thinned out, so too did they become specialized within the house. Newer designs for grand houses contained corridors specifically for movement and access and rooms became more and more specified for particular types of functions. Soon more modest houses contained specialized floor plans.[54]

The thinning out and eventual conceptual emptying of places occurred most dramatically in the work environments. As Figure 3.7 suggests, a basic difference between merchant and industrial capitalism is that, in the latter primarily, the tools and the places of production were owned by the capitalist rather than by the workers. Workers now had to labor away from home in a place and on machines which they did not own. This meant that the capitalist had to define, supervise, and control minute details of the work process. Workers came to the factory at specified times, they worked at specified places within the factories on specified machines for specified lengths of time. They could not leave their work stations without permission and they could not vary the pace of their work. Workers were constantly under the gaze of supervisors and the workers' bodies were little more than appendages to machines.

Since work was now separated from home and members of the family were now physically separated during times of the day, other geographically distinct institutions arose to supply old and new services for households. New territorial forms sprang up for production, consumption, and surveillance: school buildings containing classrooms and desks with assigned and ordered seating; prisons subdivided into cells and cell blocks; hospitals and asylums with wings and ordered rows of rooms and beds. Common to many of these structures was an arrangement of places that facilitated hierarchical access, supervision, and control.

The increasing division of labor and spatial segmentation of elements of life thinned out activities in space. Only one (or at most a few) minutely defined event was to occur in one carefully prepared place. In many cases the processes that were segmented and organized were so unusual that the structures designed to contain them were not suitable for anything else. When the processes were no longer housed in the building, the structure remained empty, abandoned, unless the building was drastically renovated. But most processes, although minutely subdivided, segmented, and thinned out, could be housed within a general type of architectural shell, if the interior partitions were flexible enough to allow rearrangement. The same basic structures with only slight modifications could then serve a variety of purposes. This was the case for institutions such as hospitals, asylums, schools, prisons or factories. Even apartment buildings were designed with movable interior partitions so that the occupant could compartmentalize his living space as he desired. This sense of a flexible and conceptually emptiable container was incorporated into the language of architecture itself. Architects still built buildings but they now called them volumes or spaces.

The power to create versatile architectural forms, and minutely to subdivide, organize, and reorganize every aspect within, made the built environment both a conceptually and actually emptiable and re-usable space or container. Seeing and using space as a container at the architectural level merges with the awareness of geographical space as a surface or volume in which events occur. The same sense pertains at both scales because modern society possesses the power through territorial control to repeatedly empty, fill, and rearrange events in space. It means that events and space are conceptually separable and that one is only contingently related to the other. People, things, and processes are not anchored to a place – are not essentially and necessarily of a place.

Modern society is dynamic. It coordinates vast and mobile populations, and it has made distant places accessible through uniform systems of transportation. Moreover, modern society exhalts science and technology, and these modes view space as abstract and metrical – as geometry. It is not surprising that a society's technical view of space becomes an important, if not the primary, practical view and experience of space. By no means, though, is an abstract metrical sense of space our only experience of place. It is, however, the primary component in our public view and it affects even our personal experiences of the world. The dynamics of modern society make it an effort to have things stay put, to put down roots. Viewing places as bundles of events takes time, and in our society things change rapidly. No wonder then that even the discipline of geography, which is most concerned with place and space, has collapsed the physical environment into an isotropic metrical space, has come to emphasize generic rather than specific places, and has reduced physical relations to metrical distances.

The role of territoriality in forming a sense of an abstract emptiable space is only one of the several possible territorial effects that can be expected in modern society. The theory suggests in addition that territoriality can be used by modern society to develop bureaucratic structures and to obscure sources of power. These effects and their relationships to abstract metrical space will be examined further, when we consider the development of the American territorial system and the work place.

Instead of exploring the other watershed described in the beginning of this chapter and in Figure 2.3 – that of the transition to civilization – and instead of considering the differences in territorial uses among different pre-modern civilizations, the remainder of this book will focus on questions of modernity. But it is important to remember that not every significant use of territoriality can be linked directly to changes in political economy. The connections described in Figure 3.3, which formed the basis of this chapter, suggest that several territorial effects can be expected to occur in any society; and that some modern characteristics can be found in pre-modern civilizations and vice versa. Moreover, the relationship in Figure 3.2, as well as the general interconnections in Figure 3.1, point out that even when territorial effects are linked to political economic relationships, the link is not always strong. Territoriality, especially within organizations, can possess a dynamic of its own. An organization can use territoriality and develop and change (as illustrated in Figure 2.2) even when the broader political economy of the society does not. To illustrate this complex connection between territorial effects within an organization and between the organization and the society, we will examine the modern and not so modern territorial uses of the Catholic Church over the nearly 2,000 years of its history. The Church is an archetype of organizational hierarchy and is one of the oldest and most explicit territorial organizations in history. It exemplifies both persistence and change in territorial use through succeeding social contexts.

4

The Church

The Roman Catholic Church contains complex examples of territorial effects. During its almost 2,000 year history is has formed an elaborate hierarchical territorial system that has influenced Church goals and policy. Until the end of the Middle Ages the Church was more advanced than most other institutions of its time. But since the rise of the Modern World, important segments of Church organization have been thought of as conservative and even archaic. And throughout its history, territoriality has played an important role. What has made the Church territorial and how has territoriality affected the nature of the Church?

Territories and the visible Church

The Roman Catholic Church began with a charismatic leader and a loosely organized group of believers and developed into one of the largest, most clearly articulated, and enduring hierarchical and bureaucratic organizations. Bureaucracies develop to help attain the goals of an institution. But frequently an organization's officials come to have interests of their own which diverge from those of the institution. Such officials subvert the role of bureaucracy to enhance their own offices and status. These officials' ambitions are often best served by their making others believe that their visions of hierarchy and bureaucracy are indispensable components of the organization. The history of the Roman Catholic Church has been particularly susceptible to this conflict of interests, especially because the original and overriding goals of the Church – the saving of souls and promulgation of virtue – are abstract and intangible. Critics of the Church have argued that the Catholic hierarchy has prevented the Church from fulfilling its religious missions. Church authorities do not agree. They contend that in fact its organization and hierarchy are sacred and essential parts of the Church and its mission. Still, even for the Church apologists, the union of Church ideals and organization is often an uneasy one.

The ethereal and practical interests of the Church and their interrelationships are reflected in the general notion that the Church has two natures. The first includes the abstract system of belief and values found in the scriptures which the Church purports to represent and to which those in heaven have adhered. This will be called the invisible Church. The second refers to the social institutions of the Church and encompasses its members, its officials, its rules and regulations, and its physical structures and properties. This we will term the physical or visible Church.[1] Church buildings, properties, holy places, parishes, and dioceses are elements in the visible Church. These are not simply things located in space. They are places set apart by boundaries and within which authority is exerted and access is controlled. In other words they are territories.

Types of Church territories

The Church recognizes and controls many types of territories, but we shall focus primarily on two: those that are set aside as holy places and church buildings, and those that are linked to such administrative structures of the Church as parishes, dioceses, and archdioceses. Roman Catholic holy places include the locations of miraculous events and the physical location and structure of church buildings. The buildings and their locations are considered sacred and, since at least the fourth century A.D., they have been consecrated.[2] All holy places are not equally holy or sacred to Catholics. Some churches are more sacred; in addition to being consecrated, they have been built near or upon the site of a miraculous event. And not all miracles are held in the same esteem. Generally, the Church has the highest regard for those places that have a connection to Christ and to the Apostles. At the top of the list are such sites as, in the Holy Land, the Church of the Holy Sepulchre and, in Rome, the site of St Peter's Tomb.[3]

Holy places are intertwined with Church hierarchy because, on the one hand, holier places often have higher ranking Church officials in charge of them, and, on the other, the authority of an official is in part derived from having authority over a sacred place. St Peter's Tomb at Rome gives added weight to the bishops of Rome for it makes them also the vicars of St Peter.

At smaller geographical scales, Catholic churches are subdivided into areas of varying degrees of sanctity. Although their building forms vary from longitudinal to centripetal plans, they all contain similar places graded according to sacredness (Figure 4.1). Overall we find first the sanctuary with its most sacred place, the altar, and then its space for the choir and presbyters, and second, the nave, which is the place for the congregation. Here too a relationship exists between rank in the hierarchy and geographical accessibility. During Church ceremonies, only those who are officials of the Church are to have access to the sanctuary, the altar being accessible

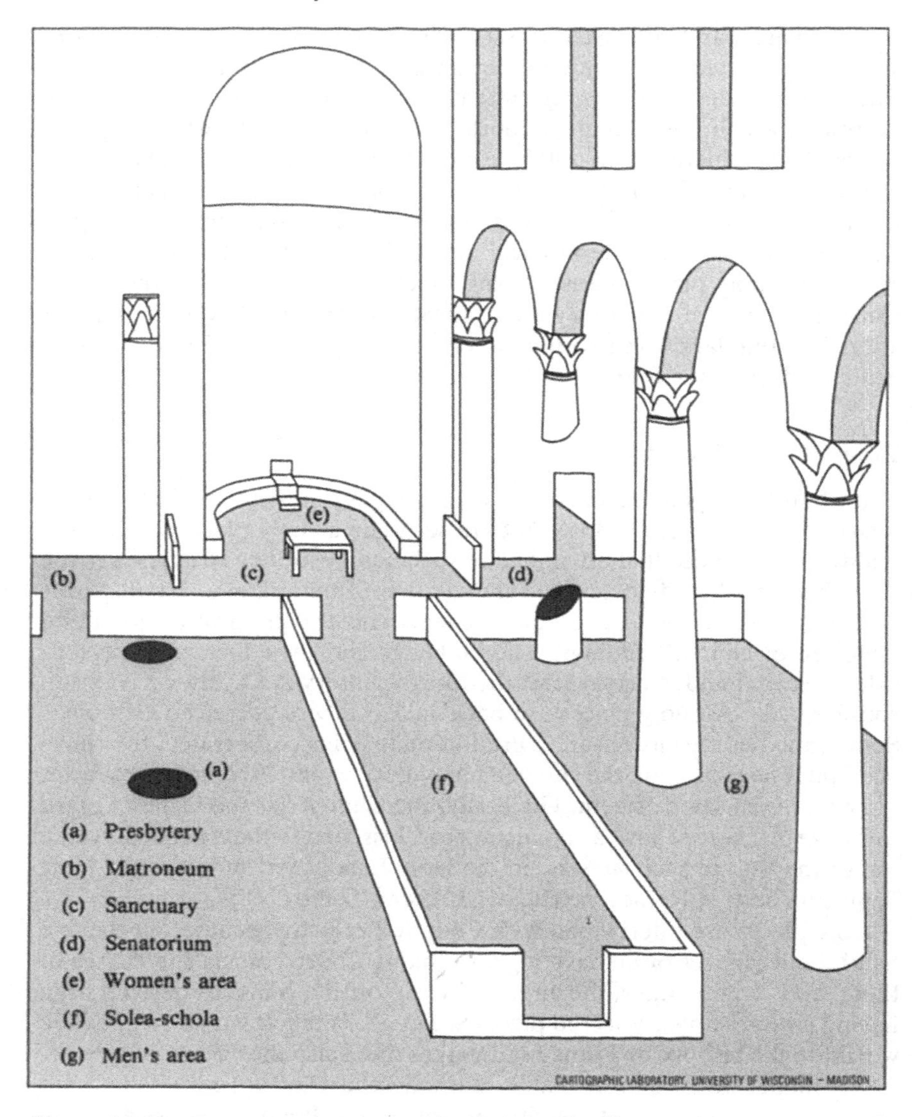

Figure 4.1 Interior areas of typical early Christian basilica
Source: New Catholic Encyclopedia. Reproduced by permission of the Catholic University of America, Washington, D.C.

only to the highest Church official, while the nave is reserved for the people.[4]

The second type of Church territory to concern us is at a geographically larger scale and refers to the units associated with Episcopal Church organization. (Although monastic systems were intermeshed with the Episcopal, it is the latter rather than the former that will be the primary focus of attention.) The Catholic Church divides most of its realm into nested territorial hierarchies of parishes, dioceses, archdioceses, and, in some areas, metropolitan sees (see Figures 4.2 and 4.3). Each of these territories is headed by a Church official whose rank in the Church government corresponds to the rank in the territorial hierarchy. Priests have jurisdictions over parishes, bishops over dioceses, archbishops over archdioceses, and the Pope over all. The relationship of course is not perfect. Not all priests have their own parishes and a bishop is the parish priest of his Cathedral's diocese.[5] Unlike church buildings and sacred places, the territorial units in the second category are not sacred. Yet they have become as important to the visible Church as are its buildings and holy places.

These two types of territories have the most to do with the internal organization of the Church as a religious organization, and they will be the focus of our attention. Their histories are important and it is their development, multiplication, and hierarchical arrangement that will be shown to be strongly associated with the development of Church organization. As critics have pointed out religion is not the only interest of the Church. The Church is also a political and economic institution. These other roles inevitably affect the functions of both types of Church territories and create other types as well. In some cases the political, economic, and religious functions of the visible Church and its territories lead to multiple and conflicting sources of control.

For example, the adoption of Christianity as a state religion by the Roman Empire not only gave Church leaders official status and hastened the bureaucratization of the Church but also placed numerous territorial units in the hands of the Church. Roman dioceses were multi-purpose territories; only one of their functions was religious.[6] Moreover, from the mid-fourth to the end of the sixth centuries the geographic limits of the Church were the practical limits of the Empire. The politics of the dissolution of the Roman Empire were far to outweigh theological-doctrinal differences as causes of the Roman Catholic–Eastern Orthodox schism. It was a political decision of the Popes to ally themselves with the West – to fill the political void left by the collapse of the Empire. Linking the Catholic Church to the fortunes of the West led the way for Church territories to become a part of feudal political economy.

A parish in the Middle Ages was often a political-administrative and economic unit as well as a religious one. Secular leaders in the Middle Ages strove to control the appointments of priests and bishops and the revenues of Church property and territory. The wealthy founded churches and gave property for their support. In return, they wanted to retain control over them. Despite these secular encroachments, the Church became an import-

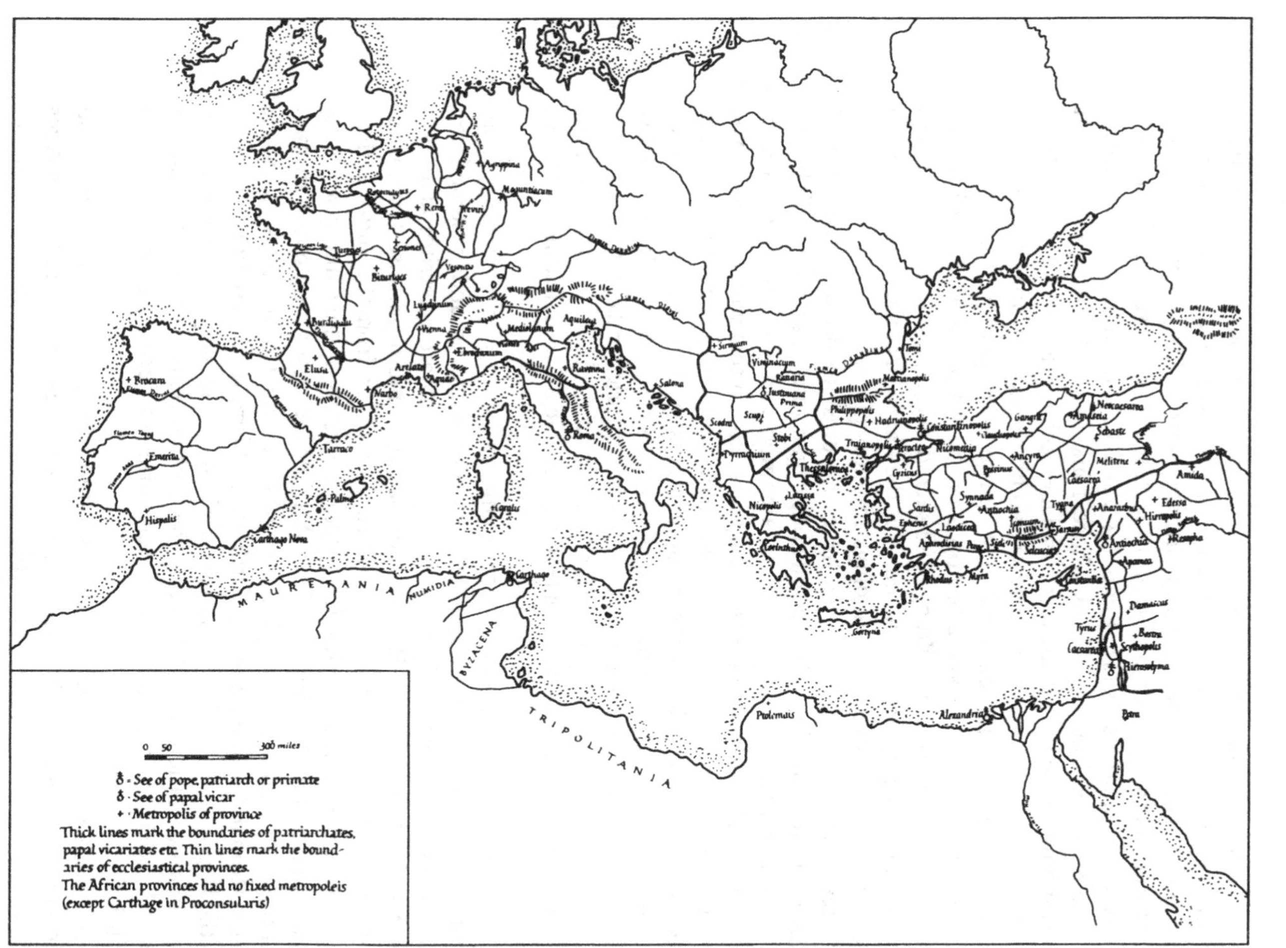

Figure 4.2 Ecclesiastical boundaries under Justinian

Source: A. H. M. Jones, *The Later Roman Empire*, *II* (Oxford, 1964).

Figure 4.3 Contemporary territorial hierarchy of Church: archdiocese of Milwaukee, WI

ant politically sovereign state with power over larger political territories, as in the formation of the Papal States which began with the donations of Pepin in the eighth century.[7] The effect of these largely political ventures and territories on the Church were extremely important. Yet much of their implication is reflected in the changing character and functions of the two types of territorial organizations to be addressed – the episcopal territories, and holy places and buildings.

The history of territoriality within the Church is complicated, even when the inquiry is restricted to only two types of territories. But the union of organization theory and territoriality discussed in Chapter 2 suggests several general territorial effects that can be anticipated even in such a complex context as Church history. Overall, we can expect the theory to guide our understanding of the Church's use of territory. We can expect that territoriality has gone hand in hand with the development of Church organization and hierarchy. As the latter increased so did the former, and conversely. Specifically, we can expect to see a positive association between development of Church territory and the sociological dimensions of specialization, standardization, formalization, and the organizational span of control. Also we can expect that territoriality has helped create some of the modern facets of Church organization such as impersonality of office and an intense territorial definition of social relationships. We can also anticipate that the internal dynamics and tipping points of territoriality, as illustrated in Figure 2.2, have influenced the development of Church structure and goals. That is, Church territoriality can be expected to have created geographical mismatches and spillovers and to have become an end in itself.

These and other territorial effects mentioned in Chapter 2 will guide our discussion of the Church. But two qualifications should be borne in mind. First, focussing on organizational and territorial changes does not mean that these features can be readily isolated and quantified. Church records are vast but incomplete, and Church historians have not assembled them in terms of the criteria of organizational theory or with an eye for territorial effects. These limitations mean that the changing relationships between territory and organization outlined above can be only suggestive.[8] Second, because we are concentrating on Church territory, and only two types at that, we are also emphasizing only a portion of the organizational character of the Church – its hierarchical organizational structure. This indeed is an important facet of the Church but we should be careful not to forget that the Church and its organization possess other characteristics as well. The clergy, for instance, see themselves as professionals, and professionals often have goals that differ from those of bureaucrats. The Church also has thought of itself as a family.[9] The bishop tends to look after his parishioners as a father does his children, nuns are 'wedded' to Christ, and clerics in the Church are often part of family in the sense that they live together and often communally.

The Church is indeed many things, not least of which is a religious body interested in ethical matters, and all of these characteristics have affected the development of Church structure. Nevertheless, the Church did emerge as one of the most extensive, long lasting, hierarchical bureaucratic organizations in the world, and its structure in turn has affected its goals. In light of the Church's other worldly aims it is all the more impressive that this institution, as opposed to other types such as armies, and governments, which have avowedly mundane concerns, became so clearly hierarchical and bureaucratic.

Primitive Christianity

Throughout most of the first 300 years of Church history, communities of Christians were small, varied in character, geographically far-flung, composed often of the 'lower and middle classes' and frequently persecuted. They existed in a society containing other religions that were hierarchically organized with clear territorial units, and in an empire that was itself a welter of organizations and territories. Yet, within this context, we find strong forces in the early Church resisting the creation of hierarchy, of a visible Church, and of territory. Even when the early Church professed the need for organization, it was often expressed in terms of the model of the family. But despite this resistance to more formal and impersonal organizations and to a clearer visible Church structure, the first 300 years of Christianity show a discernible though reluctant movement from a loosely organized collection of believers to a more formally, impersonally, and hierarchically organized group. This movement was accompanied by increases in territoriality at both scales, so that by A.D. 300, instead of a community simply residing and worshipping in a place, Christians were on the verge of using territoriality to help define their community, their places of worship, and their relationships to each other.

Jewish context

Christianity is an offshoot of Judaism. Christ and most if not all of the Apostles were Jews.[10] Most of the original Christians were Jewish converts and the original work of the Church took place in Jewish settlements. For many years it was unclear whether Christianity was but one more Jewish sect, or mystery cult, or really another religion. The link between Judaism and Christianity had a strong effect on Christian organization. Though Judaism was a state religion it spawned numerous dissident groups, one of which was primitive Christianity. Primitive Christianity was in part a rebellion against the rigid hierarchical structure of Judaism. It is significant that Judaism also had a rigid territorial structure. The link between the two is important because the earliest Christians often criticized one through the other.

Judaism contained important dilemmas which were to be echoed especially in the Christian distinction between the visible and the invisible Church. The Jews believed there was only one God and that He was ubiquitous. Yet He seemed to prefer one people – the Jews – and appeared to reside in the Holy Land and most often in the Temple.[11] The Temple was the focal point of contact between heaven and earth and its inner sanctum was the Holy of Holies. The Temple was the officially sanctioned center of religious worship for all Jews. Jews, moreover, believed that God was accessible to each man; that each could know Him through worship and obedience to His commandments. Yet they developed an elaborate priesthood which focussed on the Temple and which served as an intermediary between God and man.

The relationship between the Jews, the land, and God is complex. God set aside Israel for his Chosen People. It was a holy land, for which the Jews must make themselves deserving by following God's commandments. One means of making Jews adhere to God's will was through the development of a Church hierarchy reinforced by state authority. The closer this link, the more enmeshed becomes the territoriality of political power and religious practice. One can be used to support the other. As a state religion, pronouncements about places of worship could be enforced and concentration on one primary place could help to focus religious authority and politically concentrate and unify the state. Indeed Judaism seems to have carried this to an extreme.[12]

The Temple to the Jews was the holiest place. The Jews of Israel were expected to come there to worship and sacrifice. Even though Eretz Israel was a small country, it was difficult for Jews outside of Jerusalem to make the trip more than a very few times a year.[13] It is likely that other local places of worship arose, but there is no evidence that such places were encouraged by the established order. And assuming they were tolerated, their functions would be restricted because the Bible contains references to attempts by organized Judaism to extinguish competing places of sacrifice, though perhaps not of prayer.[14]

The origin and purpose of early synagogues is not clear. Synagogues are mentioned only at the end of the Old Testament and in the New, but were they schools, places of worship, or places of sacrifice? Part of the difficulty is that the term is Greek and seems to have meant an act, and then later a place, or assembly. The Hebrew term for synagogue is 'bet ha-knesset,' and meant originally house of gathering. An accompanying term 'bet hamidrash' meant place of study.[15] A handful of scholars believe that whatever functions it performed, the synagogue may have been an ancient institution contemporaneous with the Temple. But most suggest that it began in the Babylonian exile and continued among those Jews who were to live outside of Eretz Israel until the time of Christ when it was also found within the Holy Land. This interpretation is sensible because the diaspora Jew needed someplace in lieu of the inaccessible Temple. Although the purpose served by the synagogue before Christ remains unclear, we do know that after the destruction of the Temple in A.D. 70 synagogues were the primary focal point of Jewish social and religious life.

In addition to the Temple, Judaism had also developed a rigorous hierarchy of religious offices which were filled by select families. At the top were the Cohens or priests who conducted the services and from among whom was selected a High Priest and head of the religion. There were the Levites who assisted in the Temple, acted as gate keepers, and formed the musical accompaniment to the services. In addition there were ancillary groups providing specific services to Temple maintenance and ritual.

The building itself was in the hands of the religious hierarchy. The structure contained sub-areas, some holier than others, and access to the holiest places corresponded to one's position in the religious hierarchy. In 1 Chronicles 28 and in 2 Chronicles 3 and 4 are found the divine specifications of the first Temple edifice consisting of a vestibule and an inner sanctum: the Holy of Holies. Plans for a second Temple are found in Ezekiel 40–7 and a new Temple was built after the return from the Babylonian exile. By the time of Herod, this second Temple included a compound surrounding the original two-room design of vestibule and inner 'Holy of Holies.'

> The compound in a sense could be likened to a community, encompassing a variety of chambers which were like offices. The open space south of the Temple itself was available to Gentiles . . . There was another, smaller area which only Jewish men could enter, and yet another restricted to Priests. The inner 'holy of holies' of the Temple itself was restricted to the High Priest, and even he was permitted entrance only on the Day of Atonement.[16]

At the time of Christ, Jewish theology espoused universalism but in practice Judaism was a hierarchical organization with a hierarchical territorial structure. Jews occupied a territorial state which enforced a state religion. The Temple was a territory of worship and the Temple itself contained a hierarchy of sub-territories with the holiest place reserved only for the High Priests. In these respects Judaism was not unlike other religions at the time. Sacred places and religious hierarchy were fundamental parts of surrounding and pre-existing religions in the Middle East and were also important facets of the religion of Rome. Religious hierarchical organization, though, was only a small fraction of the hierarchy within Roman civilization.

Just as Judaism contained dilemmas, so too did early Christianity. One in particular – the material versus the spiritual – is especially important to Church organization. Early Christians were strongly eschatological; they tried vigorously not to be weighted down by officials, holy places, and territories. The New Testament contains numerous pronouncements against religious organization. Paul wrote to the Corinthians that they, themselves, were a Temple of the living God (2 Corinthians 6:16). According to Matthew, Jesus frowned upon public display of prayer; on 'those who love to pray standing in the synagogues and in the corners of the streets' (Matthew 6:5). Instead 'when thou prayest, enter into thy closet, and when thou hast shut thy door, pray to thy Father which is in secret; and thy Father which seeth in secret shall reward thee openly' (Matthew 6:6); for 'where two or three are gathered together in my name, there am I in the midst of them' (Matthew 18:20).

It is likely that such positions against organization and hierarchy were already being espoused by Jewish dissident groups, some of whom flailed against hierarchy and state religion while others wanted to purify and reform it. The Samaritans for instance had built their own temple in Mt Gerizim, and the Essenes moved from the Temple to Qumran on the Dead Sea. Here they established an alternate temple where they worshipped for the time being, hoping to return to Jerusalem when the Temple had been properly purified. The longer they remained, the more they believed that God would be attracted to their site containing a community of true believers. This idea, found in their later writings, anticipated the early Christian conceptions of the temple or the church as not a building but a community of worshippers; a social definition of territory. Furthermore, the sacrifices away from the Temple led to the foundation of the acceptance of the more general Christian notion of a symbolic sacrifice.[17] These incipient concepts were compatible with the reliance by diaspora Jews on the synagogue in lieu of the Temple as a place of worship and assembly. The synagogue was sited anywhere the faithful gathered. These dissident Jewish experiments were to form a step in the direction toward the early Christian view of a non-physical, non-territorial Christian sense of community.

Resistance to Judaic hierarchy, to state religions, and to religious territory enhanced the Christian claim to universality. It made conversion easier. It made Christian communities more flexible. But it also made Christianity more unusual and difficult for Rome to comprehend. Whereas Roman persecution of Christians may have made Christians more conscious of their own identity and increased their need for organization, the primary reason, if reason it be, why early Christians became more accepting of organization and hierarchy is that, like all groups, they needed internal discipline to continue to exist. And the New Testament could be used as easily to justify authority as it could be used to condemn it. Those in authority built upon New Testament pronouncements to justify and extend their positions. Despite the original and sincere resistance to rigid hierarchy and organization, and despite the fact that the descriptions of Church structure were originally phrased in terms of familial organization, more and more hierarchy and impersonal characteristics were soon to enter Church organization: first to define relations within communities, and second to stipulate relations among them. Accompanying these developments in organization was an ever greater territorialization of the Church.

Early Christian organization

The early Christian communities were geographically dispersed and often multi-cultural. Uniformity in their organization could not be expected in the beginning. Perhaps those communities that were originally largely Jewish might retain something of the Jewish communal structure. Jewish diasporic communities often had a council of elders to hold the community together and to enforce discipline.[18] This council (of presbyters) was often headed by a president, or, in Christian terms, a bishop. In those early churches established by the Apostles, the bishops may actually have been personally selected, but in any case the president or bishop, in conjunction with the presbyters, formed a natural governing body of elders. Yet the New Testament contains few terms of organization consistent with the old. Paul, for example, does not use the word synagogue nor archisynagōgos, nor Jewish titles for 'cleric.' And the role of women in early Christian society is far more liberal than the role accorded them in the Old Testament.[19] Whatever blend of old and new, it is clear that the Church fathers had an organization in mind – one which likely was sensitive to different regional customs and heritages of the early converts. In 1 Corinthians 12:28–30, Paul repeatedly refers to the roles of the Apostles, the prophets, and the teachers. (The list changes slightly in 1 Corinthians 12:8–10; Romans 12:6–8; and Ephesians 4:11.) The discussion is continued in 1 Clement 41–2 (written *circa* A.D. 90) wherein the Corinthians are again urged to follow the hierarchy. Ignatius' epistles (*circa* A.D. 100) to several churches address the establishment of a uniform Church government based on the role of the bishop, properly established, ordained, and assisted by deacons and presbyters.[20] In these and other contemporary documents we find concerns about heretics, about itinerant preachers who are not properly ordained entering congregations, and we find overall an urgent need for Church organization.

That the letters themselves were written to communities by highly respected churchmen who were of other communities, and that these 'outsiders' made visitations, suggest the willingness of a community and its bishop to recognize and submit to a higher authority. By the end of the second century the hierarchy both within a community and among communities was becoming solidified and explicit. This is revealed for instance in the records of council meetings. These councils were modeled after the meetings of the Apostles that took place between A.D. 50 and 52 in Jerusalem. Probably no other councils were held in the first century (there are no records of them) but there are records of several in the second century, and more in the third.[21] Disputes within and between congregations were often discussed at these local and regional assemblies and from them can be discerned the formation of Church hierarchy. For instance, the councils of Africa, held at Carthage and starting in A.D. 251, were convened by Cyprian and illustrate that there was by that time an established hierarchical connection between Church officials in Rome and in the provincial capitals. The purpose of the council was to address the question of whether a dissident Christian group within the community was correct in wresting power from their bishop Cyprian. The problem arose when Cyprian had to leave his congregation to escape Roman persecution. While away, dissident members of the congregation deposed Cyprian and elected their own bishop who then appealed to Rome for confirmation. Rome did not approve and Cyprian convened a council to decide who then was the legitimate bishop. This council at Carthage found in favor of Cyprian and asked for Rome's assent, which was given. That the decisions of this and other synods were often sent to Rome for approval indicates that even at this early period the occupant of that see was considered first among equals.

By the third century, in Weber's terms, the Church was on the verge of developing defined hierarchies of offices. Each community was beginning to have a defined sphere of competence in an administrative sense and the office holder was becoming subject to systematic discipline. But Church bureaucracy was still embryonic and such particularly modern facets as impersonality were not yet in evidence. Communities were small and scattered. The bishop was selected by the congregation and was supposed to serve for life. He and the members of the local Church knew one another. His livelihood was not derived solely from his office and though he was anointed by other bishops, he and the other Church officials did not yet need formal training to hold office. Given this emerging yet largely personal organization what can be said of its geography? Did the early Church use territoriality to enforce discipline and cohesiveness? Did the bishops use it to define their domain of responsibilities?

Territoriality

Christian communities were localized in space. Christians often lived near one another in cities, and they came together to worship. Indeed the Christians in each city tended to form a single community, and a community or city for the most part had one bishop and one governing organization. The community was geographically focussed and, in smaller ones, the bishop maintained personal authority over his assistants and parishioners. In the event that the community was large in population and area and extended beyond the city limits to the surrounding countryside, the bishop of the city allowed deacons to travel to specified places to preside over some aspects of worship. The bishop gave such concessions personally and even reluctantly and the entire congregation was still expected to assemble in the city for the important occasions and in some cases even for the eucharist, especially on feast days. As Ignatius said to the Smyrnaeans: 'let no one do anything pertaining to the Church without the bishop. Let that be considered a valid Eucharist which is celebrated by the bishop or by one whom he appoints.'[22]

Roman cities had relatively well-delimited administrative boundaries. Large cities, moreover, were often provincial capitals. The question naturally arises as to whether these early Christian communities were not only geographically localized and focussed, but whether they were also defining themselves territorially by using perhaps the city limits or the territorium of the hinterland to classify and mold the community. On first thought it would hardly be surprising if they did, for this was the general practice for defining political communities, and how else could they arrive at the neat assignment of virtually one bishop for each city? Yet the unofficial and often underground status of Christianity suggests that at this geographical scale the individual Christian communities began as non-territorial groups, and then, perhaps as early as the second century, place, if not territory, became part of the definition of the community. By the third century the group may even have used territoriality to supplement its own social definition and to help keep out strangers and to assist in enforcing discipline. Yet the territory was still primarily a unit encompassing a viable social organization and hence socially defined.

In the first century, perhaps due to eschatology and to the peripatetic character of the Apostles, the Church fathers go out of their way to make it seem that the Church visible is not anchored to territory. Early Church letters (Paul's letters to the Corinthians, Clement's first epistles to the Corinthians, the epistles of Ignatius to the Ephesians, Magnesians, Trallians, and Romans) refer to a church being at (not of) a place and sometimes soujourning at a place.[23] Although often addressing Church government, these letters do not speak directly about territory. Paul's letters to the Corinthians concern the need for Church authority and hierarchy and warnings to the congregation to beware of itinerant prophets who may beguile the congregation away from the true faith. The letters recognize that the congregation or community is in the city and that normally there is one per city, and the concept of itinerant prophets suggests that the position of a bishop or cleric had become geographically fixed. But no mention is made of using the city limits or some other specifiable boundary as a means of defining and enforcing jurisdiction over the congregation. The same can be said of Clement's and most of Ignatius' epistles. They too exhort the congregation to follow its bishops and they inveigh against false prophets, itinerant preachers, and speakers in tongues. They warn against having such people 'come among you' and they beseech the congregation 'not to admit them,' but again there is no explicit use of territory to assert control over the community. Indeed this is what could be expected, for they did not yet have the authority to back up territorial assertions.

Still the traces of the beginning of change are perhaps revealed in one of Ignatius' letters to the Smyrnaeans in which at one point he refers to himself as the bishop *of* Syria.[24] While this is but one instance, the form of address becomes more common at the end of the second century, and by the third, the connection between a bishop and a place, if not a territory, becomes explicit. Councils contain representatives from places and ask congregations to exclude heretics from their places. The example of the degree to which place, or perhaps territory, has begun to be associated with authority is seen in the controversy surrounding the bishop Origen. The theological undercurrents are complex but the geographical ones are simple. Origen was from Alexandria. He was invited by the bishops of Caesarea and Jerusalem to preach in Palestine. This angered his own bishop, Demetrius of Alexandria. Origen later returned to Jerusalem where he was ordained a priest by its bishop. This infringement by Jerusalem on what was seen by Demetrius as Alexandria's prerogative so infuriated Demetrius that he banished Origen and called two synods (A.D. 321) to censor Origen's ordination. Here clearly an administrative area or territory is linked to authority, but examples of stronger territorial controls did not, and perhaps could not, occur until the Church was no longer an underground movement.[25]

Thus over the first 300 years of Church history we find gradual and often reluctant progression from a non-territorial beginning to the emergence of territoriality of the episcopal level. The same trend can be seen at the smaller scale level of holy places and church buildings.

Many of the Apostolic fathers saw the Holy Land differently than did the Jews. Few believed Christianity to be in any sense place specific, and there is little to suggest that they thought of Israel, or even Jerusalem, as especially sacred.[26] A similar spatial detachment is found in their lack of enthusiasm for setting aside specific places for worship. We have noted their early profession that the Church is a community of believers and we have noted Christ's and the Apostles' resistance to formal edifices of prayer. God and community were inextricably linked. Life was not to be divided into secular and sacred parts. In Davies' terms, 'Worship could not be extracted from the process of living and enclosed in a secular place. Since God is omnipresent every place was holy.' According to St Daniels, 'It is not a place that is called a "church" nor a house made of stone and earth . . . What then is the Church? It is the holy assembly of those who live in righteousness.'[27] Although such sentiments can be found well into the third century, Christians came together to pray, and the convenience of meeting at a specific place gradually led to the use of that place as constituting a basic part of congregating.

Christian communities were often small, poor, and underground, and until about the mid-third century, meetings and worship took place in private houses. In the first two centuries these were usually occupied by one or more Christian families and thrown open to the congregation for religious occasions. Thereafter, it was common for an unoccupied house to be given or donated to the congregation and used solely for the purposes of meeting and prayer. Yet the houses were still 'owned' by one or more individuals because the Church was not recognized by Roman Law and could not itself own property until the early part of the fourth century.[28] Even when the meeting house was occupied by a family, there soon developed an internal differentiation of its parts, each allocated to specific functions and personages of rank. For example, a geographical differentiation of the room for eucharist is found in the following third century description:

> Appoint the places for the brethren with care and gravity. And for the presbyters let there be assigned a place in the eastern part of the house; and let the bishop's throne be set in their midst, and let the presbyters sit with him. And again, let the laymen sit in another part of the house toward the east . . . But of the deacons, let one stand always by the oblations of the eucharist; and let another stand without by the door and observe them that come in.[29]

Rooms were often modified to suit the needs of worship. In one case a room was enlarged, 'a dais was placed, probably for the bishop's chair, against the east wall, beyond which was a small vestry . . . There also appear to have been chambers for catechism and small baptisms.'[30]

By the latter part of the third century we find purpose-built churches, some of which were impressive structures; and by the beginning of the fourth, important Church business was expected to be conducted specifically within church buildings. In the synod of Cirta (A.D. 305) it was noted that the bishops met in a private house because the churches were destroyed and had not yet been restored.[31]

On the eve of its acceptance by Rome, Christianity, however reluctantly, had begun creating a visible and territorial Church. Church hierarchy had become explicit and Church officials had become to some degree separate from the rest of the community of believers. Worship was being confined more and more to the church buildings, the sacred part of life having been separated to some extent from the secular. The church building and its parts increasingly reified Church government as well as the sacred. The community itself and the authority of its bishops were coming to be more and more territorial. Bishops were of a city, and city size affected the prestige of the office.

The early Roman Church

Once the Roman Empire adopted Christianity, in the early fourth century, the entire apparatus of Roman government was at the disposal of the Church, and Church officials became part of the government. Rank and authority were more clearly specified. Bishops drew consecration fees or stipends according to the worth of their sees, grades among clergy often translated into levels of emoluments, and regular stages to priesthood were introduced. The size of the church communities and church apparatus increased to the point where, in the larger cities, the relationships between bishop and parishioners, and even bishop and church staff, were often impersonal.[32] In addition, funds became available to build churches and Christians could use former Roman basilicas and temples. Not only was worship territorialized within church buildings, but so too was the community territorialized. The power of Church officials, backed up by the state, became defined according to such territorial administrative boundaries of the Empire as dioceses. The concept of holy place in general began to assume important roles in Christian life. It was not restricted only to the church buildings, which were soon to be consecrated, but applied to places in which miracles occurred, particularly in the Holy Land. Some still resisted these changes,[33] but the general direction was toward greater hierarchical centralization, differentiation, and territoriality.

The richest single source illustrating the link between bureaucratization and territoriality is the accumulated record of the Church councils and the canons they promulgated. The number of these councils increased enormously once the Church was recognized by Rome. We noted that councils were called by bishops or metropolitans of major cities. This custom continued, but now the format of meetings was modeled after the legal political system of the Roman Senate and indeed the Emperors convoked and presided over many of the important ones. These were called ecumenical councils. Religion was the primary interest of the Church but matters of faith were, from the very beginning, inseparable from issues of organization, which explains why so much of the business of the councils concerned organization.

The start of the fourth century, when Christianity became the official religion of the Empire, to the Roman Collapse in the fifth, defines the period during which the Roman Empire helped organize and consolidate Christianity. The Church records show there to have been hundreds of regional councils and four ecumenical ones during this period. Even an abbreviated list of all the canons these councils promulgated would take volumes. A better use of space here would be to summarize the results of two early councils, Nicea and Antioch, in order to give the flavor of these meetings and the attention given to organization, and then to sample from among the rest, and abstract those canons concerning territoriality and hierarchy. These will be discussed according to which tendencies and combinations of tendencies of territoriality they exemplify.

The first and extremely important ecumenical council, Nicea, was convened in A.D. 325 by the Emperor Constantine to address the Arian heresy, named after Arius, who claimed that the Son was not the equal of the Father. The heresy was rejected by the council and a 'true' doctrine was set forth in the Nicene Creed. This declared that the Father and the Son share the same divine essence. After this was resolved (and an official time for Easter established), and Arius was exiled by Constantine for five years, the council addressed matters of Church discipline in twenty canons. These included rules: forbidding the elections of neophytes to the priesthood or episcopate; excluding self-inflicted eunuchs from holy orders; rescinding improper ordination of bishops and priests; declaring that those who have left the Church because of persecution shall be readmitted after twelve years of penance; forbidding denial of communion to any one likely to die; forbidding bishops, priests, and deacons from having women, other than their close relatives, in their houses; deposing clerks guilty of usury; and ordering that prayers during Sundays and Pentecost should be said standing.[34]

In addition, among these canons several dealt explicitly with territory and place. Canon four declares that a bishop ought to be consecrated by all of the bishops of the province, but if that is impossible, then by at least three with the written consent of the absent bishops and later confirmation of the metropolitan. Canon five orders that he who has been excommunicated by his bishop shall be denied communion by other bishops. Canon fifteen forbids bishops, deacons, and priests to leave their own cities (congregations) for another. If they do they are to be forced back. Canon sixteen is similar to fifteen except that it is addressed to the bishops of other congregations not to receive or entice away priests or deacons from another congregation. If a bishop ordained a man belonging to another church, the ordination is invalid. Canons six and seven refer to jurisdictional and territorial disputes between bishops and metropolitans or archbishops. The authority of the archbishop or metropolitan was reaffirmed and some of the hierarchy among metropolitans was stipulated. Canon eighteen refers to the internal territory of church buildings. It forbids deacons to give the eucharist to priests, to receive it before priests, and to sit among them.

The council of Antioch, A.D. 341, is important for its mixture of politics and theology and for its territorial canons. (There is disagreement about whether these were actually promulgated in 341 or before.) The council was commenced to dedicate the Golden Church of Antioch. Over ninety bishops attended, none of them was from the West and many were Arians. Eusebius of Nicomedia was the principal bishop. He had usurped the see of Constantinople, and was, among other things, held responsible for murders that occurred when he was present in Alexandria. He was condemned and replaced by Gregory, an Arian. In addition to engaging in intrigue, the council formulated creeds concerning the Son and the Father and enacted twenty-five canons. Number three forbids priests and deacons to absent themselves for long from their dioceses. Number five states that if any priest or deacon has opposed his own bishop and has separated himself from the church and collected a private congregation and disobeyed the summonses of his bishop, he shall be deposed. Number six forbids bishops from receiving any one excommunicated by another bishop. Number nine orders all bishops of a province to obey the metropolitan and to give him precedence. Number eleven states that no bishop or priest may go to the Emperor without the consent in writing of the bishop of his province. Number thirteen deposes a bishop who ordains uninvited in another province. Number fifteen allows no appeal from the unanimous decision of the provincial synods. (On the face of it this makes the hierarchy stop to a great extent at the level of the archbishop. He is virtually autonomous. Yet approaches to Rome were allowed on occasion and number fifteen in effect discourages such approaches.)[35] Number sixteen states that a bishop who is not chosen in a regular synod – one which has no metropolitan – is to be deposed even if elected by the people of his diocese. Number twenty-one forbids translations of bishops from one see to another and number twenty-two forbids bishops from interfering in the see of another bishop.

As these councils reveal, once Christianity had Roman legal authority behind it, a dramatic acceleration occurred in the definition of Church hierarchy and its use of territoriality. Numerous other councils and their canons also reveal that authority within the hierarchy was intimately related to territory. The theory of territoriality can help make clear why this would be the case. A most basic consideration is that the Empire had a vast population, geographically dispersed and with varying degrees of mobility, and such conditions would make territoriality the simplest and clearest means of defining and dividing populations into groups and of assigning them to the supervision of Church officials. Indeed the territorial boundaries, for the most part, were already clearly demarcated as political jurisdictions and generally recognized by the population. The Church could simply use them as a mold for its own authority. By the late fourth century urban Christian communities in many cases had grown to the point where personal knowledge of congregants and informal peer pressure would not suffice to define a community or enforce discipline. But effective authority could be had by enforcing territorial assertions of control.

The Church did not control the residence of the parishioners though some canons did affect long-range movements. For example Arles, A.D. 314, number sixteen (which appears to be the same as Elvira, A.D. 305–6, number fifty-three), and Antioch, A.D. 341, number six, in ordering that for one who is excommunicated to receive communion again, he must receive communion at the same place in which he was excommunicated, asks that the person be in, or return to, a territorial jurisdiction. For the most part the canons used an individual's place of residence within a church territory as a means of assigning that person to a particular set of clergy and to a particular church, as in Nicea, number five, wherein those denied communion in one place may not receive it in another. The Church, however, could and did attempt to control the geographic location of its officials and used territoriality to keep them in place, to define and delimit their authority, define channels of communication, and to circumscribe hierarchically Church responsibilities. But in so doing the Church left itself open for other territorial effects such as the possibility of mismatches and spillovers of authority and territory, the possibility of having territory become an end rather than a means, the possibility of having territory create inequalities in access to resources and authority.

Turning to the general association between hierarchy and territory, it is striking how much attention was paid in the canons to defining and delimiting the responsibilities among archbishop, bishop, priests and other Church officials by using territorial circumscriptions of authority. We find general stipulations that all clergy, including monks, and the church buildings they occupy, are under the authority of the bishop of that jurisdiction (Chalcedon, A.D. 451, canons four and five); that all disputes within a diocese be settled by the bishop (Chalcedon, A.D. 451, canon nine) and if still not resolved that they go to the archbishop or metropolitan and not to another province, and moreover that all clergy ordained in a community remain in that community (Arles, A.D. 314, canon two). The canons indeed make the archbishop extremely powerful and more or less autonomous within his own see. His authority is difficult though not impossible to circumvent as neither bishop nor priest may go to the Emperor for appeal without the archbishop's consent (Antioch, A.D. 341, canons eleven and fifteen). The archbishop should be obeyed (Antioch, A.D. 341, canon nine) and should be involved in the appointment of bishops within his province. If for some reason he is unable to perform this function at least three other bishops within the archdiocese are to represent him (Nicea, A.D. 325, canon four; Arles, A.D. 314, canon twenty).

Church hierarchy exists beyond the metropolitan or archbishop though it is not at this period fully worked out. There are, however, references in the councils to the order of metropolitans, usually referring to Rome first (Nicea, A.D. 325, canons six and seven; Chalcedon, A.D. 451, canon twenty-eight), which corresponded to the authority that the Church of Rome actually exercised by this time.

Since territory was intimately connected with the definition of authority, a prime means for Church officials to resist authority was to leave their territory. Evidence for this practice are the numerous prohibitions against it. For instance there are rules against a bishop leaving his diocese without permission and often separate canons specifying what the consequences would be (Nicea, A.D. 325, canon fifteen; Arles, A.D. 314, canons two and seventeen; Antioch, A.D. 341, canons thirteen, twenty-one, and twenty-two; and Sardica, A.D. 347, canon one). There are prohibitions against deacons and presbyters leaving the diocese or parish of their church without permission (Arles, A.D. 314, canons two and twenty-one; Laodicea, between A.D. 341–381, canons forty-one and forty-two). There are prohibitions against a bishop receiving or enticing away presbyters, deacons, and others, and sometimes separate canons specifying the consequences (Nicea, A.D. 325, canon sixteen; Antioch, A.D. 341, canons six and thirteen; Arles, A.D. 314, canon seventeen; Sardica, A.D. 347, canon nineteen; Chalcedon, A.D. 451, canon twenty).

Such illegal moves of Church officials revealed mismatches and spillovers between territorial authority and the individuals and relationships to be controlled. The only means of preventing such mismatches was to have larger territorial authorities enforce prohibitions against such movements and coordinate the activities of smaller territories. In this respect whereas territoriality offers advantages to hierarchy it also has a momentum of its own to increase the need for more hierarchy and bureaucracy while diminishing its own effectiveness.

Perhaps the most obvious territorial friction to the smooth functioning of bureaucracy comes from the fact that territoriality also has the tendency to be used to create inequalities among units and thus among the officials who govern them. Although treated in many respects as equivalent levels within a hierarchy, each see was not in fact equal in wealth and prestige. The warnings against translating from one see to another, as expressed by Bishop Hosius, 'the cause . . . being well understood; for it had never been seen that a bishop left a large bishopric to take a lesser one . . . ',[36] and the warnings in council of the fourth synod at Carthage, *circa* A.D. 398, canon twenty-seven, stating that neither a bishop nor any other ecclesiastic shall go from a smaller to a more important place without written consent of his superior, clearly reveal that unequal wealth of territorial units was indeed a factor.

Coupling hierarchy with territoriality aggravated vertical inequalities. Archbishops command the resources of their provinces. They are also responsible for Church discipline and doctrine and it was they who convened and presided over the councils wherein long-range doctrines and policies are prepared. Being an archbishop gave one these privileges and being the archbishops of Constantinople or of Rome gave one even more. It is not surprising that the canons reveal that control over a specific territory was often regarded as an end rather than a means to an end, and that conflicts over theological matters and even over personalities were often displaced as conflicts among territories.

Territoriality cannot only clarify relationships of authority, it can make them impersonal. Whoever is to assume an archbishopric is to have control over the entire Church hierarchy within that territory. The pronouncement of Chalcedon, A.D. 451, that the bishop control all monks and new buildings within the territory (canon four), and that clergy of chapels and monasteries are to submit to their bishop (canon eight), have an impersonal ring. So too do those canons stipulating the relationship of the parishioners to the bishop. Arles, A.D. 314, canon sixteen, for example, states that those excommunicants will receive communion in the same place they were excommunicated. These and other canons point clearly to the potential of using territoriality as an impersonal mold. New parishoners can be molded to a church community by virtue of their location, and priests, bishops, and other officials can be linked to each other and to the community by location. The geographic dimensions of any new rules are molded by the territorial boundaries of the communities. Clearly, even at this early stage, the impersonal potential is there. But we can surmise that in only the very large urban centers, where Church officials numbered in the hundreds and congregants in the thousands, did this potential have a significant impact. Moreover, after the dissolution of the Empire, this potential was overshadowed and did not again become important until the eleventh and twelfth centuries.

.The use of territory seems also to engender more territory in the sense that territorial definitions of authority become basic mechanisms for Church organization, and also in the sense that creating new dioceses and parishes becomes a means of enhancing a bishop's or archbishop's standing. Chalcedon, A.D. 451, canon twelve, forbids any bishop from dividing his province by obtaining letters patent from the Emperor, and Sardica, A.D. 347, canon six, forbids consecration of a bishop for a small place when a priest suffices. (See also Laodicea, A.D. 341–81, canon fifty-seven.)

So many of the canons are rules about the conduct of parishes, dioceses, or archdioceses that the people to whom they apply may have been overlooked. It was common practice in some cases for rules to be promulgated for classes of territories. Territory can be used to draw attention away from the sources of conflict, whether they be theological or personal, and territory can make it appear as though the problem is a conflict between places. The see of Constantinople vies with the see of Rome, Alexandria with Jerusalem and Antioch. The see becomes an object of pride. In combination, then, these tendencies can indeed make the territory appear to be an end rather than a means.

From the evidence of the canons it does not appear that the Church saw its episcopal territories as conceptually emptiable and fillable spaces. The closest they came to an empty space conception may have been when the Church decreed in the twelfth century that parishes should be erected wherever there were none. Another tendency, though, does come to the fore in these documents. Reification is occurring throughout the entire process of making the Church visible. Allusions to this tendency are found whenever the congregation, the parish, and the diocese are referred to as the physical Church. The parish is to this day spoken of as a reification of the Christian community.[37] But the most abundant references to reification concern the level of church buildings. As noted before, by the third century A.D., the church building was already becoming a sanctified place containing a hierarchy of sites within it that were accessible to different levels of the Church hierarchy. Gangra, A.D. 325, canon five, makes anathema those who despise the house of God, and canon six disallows private religious meeting outside the church, without the consent of the bishop. Nicene, A.D. 325, canon eighteen, forbids deacons to 'sit among the priests.' In Laodicea, A.D. 341–81, canon six, we find that no heretics are permitted to set foot in the church, in canon fifteen we find that only those appointed may ascend the pulpit, and canon nineteen directs that only clerics shall be permitted to approach the altar and communicate. Deacons and members of other lower orders may not sit in the presence of a priest, unless bidden to do so by the priest (canon twenty). The so called 'agape' shall not be held in the church and no one shall place couches in the houses of God (canon twenty-eight). Women may not approach near the altar (canon forty-four); and the priests shall not enter and take their seats near the altar before the entrance of the bishop (canon fifty-six).

Moreover, clerics are prevented from being in other places and undertaking specific activities elsewhere. They may not sacrifice outside the church, they may not enter public houses, or perform religious functions in heathen cemeteries. Further stipulations consecrating and subdividing the use of the church building were to continue to the end of the Middle Ages.[38]

Reification is very closely linked to the process of displacement. In some cases it is impossible to draw the line between them. Generally, reification makes visible the agent influencing or controlling. It brings it down to earth. The church building, as well as the parish and diocese, reminds us, through reification, of God and the Christian community. Displacement goes further in that it makes us lose sight of the fact that territory is a device or instrument and instead leads people to believe that in some sense territory itself is the object represented. The buildings as consecrated places can actually be perceived as having powers themselves. The power is enhanced when the site is especially sacred and when the building contains relics. Pilgrimages are made to a place because the place itself can bring comfort or heal. To damage a consecrated building, or to enter a sacred place improperly, can do you harm. Reification and displacement make places themselves appear to have power.

The early Middle Ages

With Constantine's conversion (*circa* A.D. 313), the Church had received the protection and stimulation of the Empire. Indeed the influence of the Empire on the Church was far greater than the Church on the Empire. As long as there was a Christian Emperor the Church could be uniform and organized, but enormously dependent on secular power. The period around A.D. 500 marks the beginning of an abatement, and for several hundred years a retreat, from the development of increased uniformity of Church rules.[39]

When the Western Empire crumbled, the Church was left as a far-flung religious organization without an equivalent, large, secular organization to support it. The Church could not itself fill the political void, so it sought to make alliances with new and emerging powers in the West that had the potential of becoming great powers. The Germanic invaders, and their heirs, the feudal princes, conceived of their position as divinely anointed priest-kings. They exercised strong control over the Churches in their lands and transformed their religious councils into national assemblies (attended by secular lords as well as by Church bishops) over which they presided.[40] The alliance of the Western Church with these secular powers widened the gap between the East and the West, and increased further the Church's dependence on newly emerging political forces. Except for Charlemagne's Empire, these secular rulers did not develop a substantial unified political system. Rather political entities were fragmented into a complex and ever changing web of small kingdoms which tended to divide the Church into a series of petty national and local units. These Dark Ages did not produce ecumenical councils and grand syntheses. Rather the councils were local and many of the canons reveal the degree to which the precursors to feudalism had eroded hierarchical authority and territorial control. This 'feudalization' of the Church came to be reversed around the twelfth century which marks the beginning of the end of feudalism in general and a resurgence in Church hierarchy and organization.

Feudalization

From the end of the sixth to the eleventh century we find a progressive enfeudation of the Catholic Church which considerably diminished its hierarchical organization and territorial authority. At the bottom of the Church hierarchy the effects of enfeudation can be seen in the rise of private Churches over which the official Church had little control. The occasional Roman practice of a wealthy family endowing a chapel or church and selecting its priest became rampant after the break-up of the Empire. Indeed the Church early on recognized and sanctioned its practice. Orleans, A.D. 541 canon thirty-three, formally required that anyone who wished to found a parish must provide for its support with property and clergy. Many who built and endowed these private churches also wanted for themselves and their heirs the right to appoint the clergy. These and other privileges were to be partially conceded to the founders while the Church still tried to retain some degree of control.

Church hierarchy attempted to maintain some authority by requiring that clerical appointments be approved by the bishop and that the bishop still be responsible for consecrating priests and church buildings (Orange, A.D. 441, canon ten). But if the bishop did not accept the founder's choices, he was obliged to find one that met with the approval of the founder (Toledo, A.D. 655, canon two). The Church not only conceded much of the appointments to the founder but also permitted him to alienate the property (Frankfort, A.D. 794). Even the right of patronage was alienable. 'In many cases the private Church was actually transferred from the founder's family and from later proprietors of the domain on which it stood, and by a commercial transaction came into the patronage of persons who lived at a distance from it and had no special reason to take an interest in the worshippers.'[41]

The problem of control and territory was complicated even further by the fact that bishops were often founders and proprietors of such churches in dioceses that were not their own (Orange, A.D. 441, canon ten). Even those parish churches which were not private and were part of the diocesan organization frequently became the equivalent of private churches because the resident priests needed the physical protection of the local nobility. In effect such churches were enfiefed to the local lord.

Similar transformations occurred higher up the Church ladder. The bishops and archbishops were not only religious heads of vast areas but also lords of estates. Peasants and nobility were often vassals of bishops and archbishops who were capable of raising armies and maintaining law and order. But these prelates in turn were vassals of higher nobility and kings. (In the tenth century Otto I of Germany made courts of numerous bishops and thus his direct vassals.) Lay leaders had many reasons to be interested in who was appointed to the sees within their domains and took great pains to influence such appointments. By and large these efforts were successful.

The process of enfeudation weakened Church hierarchy and bureaucracy as well as its territorial rigidity. The loss of control by Rome meant both a gain of autonomy on the part of local units and a gain of control over these units by secular authorities. In both cases local territorial control weakened the hold of Rome.

The increasing role of secular authority in Church affairs did not, however, lessen the importance of the Church as a religious institution in the eyes of contemporaries. On the contrary, it suggests that the sacred and the secular could not be separated, that the sacred is present in all realms of life. The permeation of the local community by religion and Church affairs is most clearly seen in the role of the church buildings themselves. Church law remained unyielding in its view that the church was a consecrated place, that worship should occur within its walls, and that the church territory should be divided into more and less sacred parts to which members of the community would have differential access. But the environment of the local community made such sharp distinctions of use difficult to enforce. During the early

Middle Ages, the church building was often the single largest and most secure structure in a community and it became the custom for parts of the building to be put to non-religious uses. In times of political instability, the church served as a place of defense. At other times it was a meeting place for village affairs, a town hall, a hospital, and inn.[42]

The late Middle Ages and Renaissance

Enfeudation began to abate and even be reversed in the twelfth and thirteenth centuries. The causes are far too complex to outline here, but some of the significant effects of the changes were these. The relatively closed and insular political economic units of feudalism were being transformed into a more open and interdependent economy. The role of superstition if not religion in everyday life was pushed back somewhat by an increase in secularism and empiricism. The King became less of an anointed person and more of a secular ruler. In many respects, society in general was more willing to have the Church manage religious functions, and the Church actively sought this role. But more and more the term Church came to mean those who were ordained by the Church hierarchy, not those who were Church members in the broadest sense.[43] The Church's attempt to re-establish Church control over its own hierarchy led to a narrowing of the religious domain. Ecumenical councils were convened, canon law became codified and extended. Church discipline for the most part increased, and the Pope became the undeniable focus of Church leadership.

A general strategy accompanying the centralization of the Church and its attempt to regain control was to reassert territoriality and to decrease the possibility of secular forces taking over Church territory. This took place by reiterating claims laid down in the canons of earlier periods. The power of bishops is reasserted over those in his territory. In the Lateran council, A.D. 1123, canon nineteen ordered that monasteries recognize their bishops and in Lateran, A.D. 1215, chapter nine, it was ordered that 'bishops provide that all within their dioceses use the same rites.' In order to prevent mismatches, spillovers, and the use of territory as an end, familiar prohibitions against pluralities and translations were promulgated (Lateran, A.D. 1179, canons thirteen and fourteen; Lateran, A.D. 1215, chapter twenty-nine; and Lyons, A.D. 1274, canon eighteen). Similarly we find the prohibitions of bishops interfering within jurisdictions of others, as in Lateran, A.D. 1139, canon three (forbidding bishops from receiving those who have been excommunicated by another bishop), and Lyons, A.D. 1274, canon fifteen (forbidding bishops from ordaining those who belong to another Church). Also enacted were familiar restrictions on the power of bishops when on visitations (Lateran, A.D. 1215, chapters thirty-three and thirty-four).

New prohibitions were in place to prevent the control of Church offices by investiture and to strengthen the power of the Pope. These made Church offices more bureaucratic and hierarchical. Rome, A.D. 1059, canon six, forbids priests and other clerics to receive private churches. In Clermont, A.D. 1095, canon five forbids the 'appointment of laymen, and everyone under the order of subdeacon, to bishoprics'; canon six forbids the purchase of a benefice; canons fifteen and sixteen 'forbid clergy to receive any ecclesiastic preferments from laymen or for them to make any such investitures'; and canon eighteen 'forbids the laity to have chaplains who are not under the authority of a bishop.' These were repeated and elaborated in later councils and formed part of the Churches' position on investiture. In Lateran, A.D. 1112, Pope Pascal revoked the right of investiture. Lateran, A.D. 1123, canon twenty-two, declares null and void all orders conferred by inappropriately ordained bishops, canon fourteen forbids laity from interfering with Church property. Lateran, A.D. 1139, orders lay persons who possess Church properties to restore them to the bishop, and Dalmatia, A.D. 1199, canon eight, condemns lay patronage of Church officials.

Arresting power from the secular and increasing the bureaucracy of Church officialdom focussed Church authority at the top. The Pope moved from a position among the bishops of *primus inter parus* to absolute sovereign over the Church. One of the most important justifications for the Pope's central role was the 'spurious' Donation of Constantine. This document was interpreted to mean that the Pope alone can depose and restore bishops; that he alone can make new laws, set up bishoprics and divide old ones. He alone can translate bishops, and his legates have precedence over all bishops.[44] By 1335, the Pope exercised his right of appointment and bypassed local elections of clerics.[45] Whereas this right of appointment appeared to make the power of the Pope absolute, in many respects it weakened his effective control over the Church in the long run. First he could not possibly know all the people he was appointing, and they more often than not were sponsored by local forces already in power. Second, the princes found a centralized Church power in Rome easier to deal with (or ignore) than a multitude of different sources.

During this period little change in official attitude occurred towards the church as a building (although many of the heresies were hostile to church buildings). It remained first and foremost a consecrated place which had internal divisions of sanctity from porch to altar. The old rules prescribing proper conduct in the sub-areas within the church building remained applicable, but so too did the custom of using the church for secular purposes. Despite the fact that the latter part of the Middle Ages was a slightly more secular time, and that the Church itself was ready to increase the distance between the secular and the sacred, the church building often remained the most important physical structure in the community and in its use did not reflect sharp distinctions between sacred and secular. Rather it continued to serve, among other functions, as a town hall, a place of defense, a place of refuge, and a hospital.

The separation of the Church from the rest of society is perhaps more clearly illustrated in two non-territorial, but yet spatial examples. The first concerns the position of the priest during the service. It was the custom until the twelfth century for the priest to face the congregation during services; but the Gregorian reforms of the late eleventh and twelfth centuries had the priest now face the altar with his back to the congregation, thereby increasing the 'personal distance' between priest and parishioners. The second concerns the accessibility of the Bible itself. By the ninth century the Church physically monopolized possession of the Bible declaring that only clergy should possess it. In 1080 the Church forbade any layman from even reading the Bible in its entirety. 'From the Waldenses onward attempts to scrutinize the Bible became proof presumptive of heresy.'[46]

From the twelfth century on, the Church made a concerted effort to re-establish and extend its formal organization and hierarchy. This was undertaken in large measure by re-establishing the link among hierarchy, power, and territoriality, and by reducing both local autonomy of territorial units and the control over them by secular power. Within the Church, the impersonality of offices increased, and canon laws were expanded and codified, organization was centralized, and hierarchical power and territoriality were made explicit. But by the fourteenth century the machinery of government, its rules, regulations, and size, began to take their toll. There was corruption, inefficiency, and blatant prostitution of values, all of which were manifested in the Church's use of territory. Differential territorial access to resources created enormous inequalities in the Church hierarchy and between the Church and secular world. Control of territory and the appointment to a particular see often became an end in itself. This value of territorial control increased the numbers of translations and pluralities (or mismatches) between individuals and territories. Moreover, the Church's efforts at consolidation made it both more separate from and more visible to the community. As unambiguous leader of the Church, the Pope came to stand for the Church itself.

The end of the fourteenth century also saw increasing social unrest and general resentment towards authority which eventuated in the Reformation. Profound economic changes were taking place. The number of rootless people increased and cities were swelling in size. Secular interests tended to balk at some of the Church restrictions on business and usury. The separation and even aloofness of the Church was part of its general disengagement from current political economic change. As a conservative landlord and property owner, the Church did not lend a sympathetic ear to the needs of the newly emerging capitalist interests. Capitalism, as Weber pointed out, did not thrive in places dominated by the Catholic Church. Many heads of government who allied themselves with capitalist interests had an eye to appropriating or 'nationalizing' Church property. And corruption in the Church was widespread. The hierarchical and territorial consolidation of Church power and its increasing visibility gave the socially dissatisfied a clear objective. If Church corruption was not greater than ever (and it probably was), the source seemed to be more readily identifiable. It was the visible facets of the Church the early reformers attacked.

The Reformation and after

Important changes have taken place in the Catholic Church since the Reformation but in terms of territoriality and hierarchy these were minor when compared with the variety of organizational structures that were to come about within the Protestant denominations. The structures of these denominations demonstrate again that territoriality is associated with organizational hierarchy and bureaucracy. Moreover, the establishment of the earliest Protestant Churches repeat, though in a more compressed span of time, earlier Christian dilemmas concerning Church hierarchy, territory, and religious purpose. Despite the fact that Protestantism offers a wider range of alternatives to the questions of Church structure and faith than did Catholicism, it is important to recognize that Protestantism encountered problems concerning authority and faith that were similar to those we described for early Catholicism, and that territoriality again played a predictable role. We will outline some of these issues regarding Protestant Church structure and territory before summarizing some of the post-Reformation changes in Catholic Church organization.

Protestantism

Protestant reformers attacked the more visible elements of the Catholic Church: the Church's hierarchy of officials, its rituals, and its physical monuments. The church building itself, because it is consecrated by Church tradition, was singled out as the major objectionable territorial entity. For the most part the geographical parishes or dioceses were not mentioned. As we noted, attacking the consecration of buildings, and prayer in buildings, was not new in Christianity. It is found in the Old and New Testaments. It is found in the fourth-century council of Gangra by the followers of Erustatius, and again on the eve of the Reformation in the movements of the Alumbrados or Illuminati, and the Anabaptists.

These groups selected the visible facets of the Church because these aspects were thought to have drawn attention away from Christianity's true objectives. Reformation leaders held that the Church hierarchy subverted the fundamental principles contained in the scripture and that the only means of regaining them was to purge the Church of its organization and to return to the conditions of primitive Christianity. In attacking hierarchy and the visible Church, Protestantism, as with primitive Christianity, was objecting to the separation of the sacred from the secular. Instead of occupying only a segment of life, Protestant discipline was to penetrate every facet of life. Protestantism's emphasis on 'wealth as duty' and on 'a calling' made this movement of vital interest to capitalism. The link between capitalism and Protestantism was strengthened by its close political ties to emerging capitalist countries. Protestantism was to become the state religion of the most advanced capitalist nations in Europe.

The conditions of primitive Christianity which Luther desired to re-establish were defined by him in extremely personal and idealistic terms as being based primarily on faith.[47] Luther attacked the distinctions between priests and laymen.[48] He condemned those who found religion in food and vestments in holy places and on holy days.[49] Sometimes he allowed that churches were necessary simply as convenient places of worship: 'We should not disallow the building of suitable churches and their adornment, we cannot do without them. And public worship ought rightly to be conducted in the finest way. But there should be a limit to this, and we should take care that the appurtenances of worship be pure, rather than costly.'[50] But at other times he would have them eliminated altogether: 'The Church is not bound to any one city, person or time.'[51] He saw the physical form as displacing attention from the sacred: 'we are diverted from God's commandments by such a stir and clamor.'[52]

Needless to say a community of faith is not easy to establish. A religious organization needs more than this for direction, especially because, like the primitive Christians before them, the Protestants were not operating in a social vacuum. They were competing with the Catholic Church, with a soon burgeoning group of Protestant faiths, and perhaps most significantly, they were imbedded in political institutions which were going to continue to use Church organizations to supplement their control. In Europe, Church and state were closely linked, as well they should have been from contemporary perspectives, for saving souls was still serious business. A religious organization could not expect to take hold without some political support. And political interests saw new religious groups as a means to political ends.

Luther's community of faith would need a more visible organization whether Luther wanted it or not, and the success of his movement forced him to address organizational issues. His writings suggest that he preferred a form wherein the congregation or Christian Assembly has the right to judge, appoint, and dismiss teachers.[53] He noted that even the primitive Church had discipline and if the present Church cannot enforce its rules then the civil authorities should. This did not mean that one group was set apart and over another, because the civil authorities were also Christian. It meant the Church needed civil authority to help it to reform.[54] Calling on the secular authorities for support and enforcement of Church conduct was to forge a close link between Church and state. In several countries Lutheranism became a state religion.

Luther was either against church buildings entirely or tolerated them as convenient, but unconsecrated places for assembly and worship. Yet tolerating churches meant tolerating or encouraging worship in them and this raised the problem of deciding what other functions they were allowed to contain. This was not an insurmountable issue in the early Protestant communities because in binding religious to secular life, the community placed severe strictures on the latter. Several early Protestant reformers tried to ban the reading of anything other than the Bible and the singing of anything other than hymns, so that permitting only worship in a building may not have made what happened inside the building very different from what happened outside.

At the episcopal level, Luther accepted, without question, the use of the parish and diocese. This was understandable because the parish had become also a secular unit of administration and if Lutheranism was to receive state support, such governments would likely use existing political territories to define and mold the religious community.[55] Luther insisted on mandatory church attendance within the parish even for unbelievers, to be enforced by the local secular authorities;[56] 'whoever is in the parish is of the parish.'[57]

The parish continued to be a political–administrative as well as a church unit. In Lutheran Sweden the former Catholic hierarchy and episcopal territorial structure were retained virtually intact. The only significant change was in liturgy and in procedures of appointment. The old parish structure remained and became the unit of census. The parish priest recorded all major events such as births, deaths, land holdings, and value of farms. If anyone entered the parish he was to register with the parish priest. When the Swedish government officially recognized the rights of other religions, the parish structure pertained to them as well. A non-Lutheran area would have a non-Lutheran church which would have the responsibility to keep parish records.

Whereas Luther did not dwell much on Church government (although from the beginning Lutheranism was firmly enmeshed in government hierarchy and Lutheran churches maintained Catholic episcopal forms), Calvin had much more to say about Church government, and early Calvinist movements 'co-existed' with such governments rather than becoming a part of them. Calvin's attack on the Catholic Church was similar to Luther's, but Calvin's interest in Church government and discipline came from his belief that faith was not enough to define community. Because Calvin saw man to be a greater sinner than did Luther, Calvin believed man needed more guidance and supervision. Man had to strive to do good, but could not succeed without the aid of the Church. To a Calvinist, faith was a gift of God, given to a select few who had been predestined for salvation. But even they were not perfect in their faith, and no outward signs reliably distinguished between these and the reprobates. The Church, of necessity, included both, and both needed the guidance and discipline of the Church.

Calvin's Church government was designed to work with secular government, and in particular the government of the city of Geneva. The Church was to have four layers of officials: the teaching doctors, the preaching pastors, the disciplining elders, and the administrative deacons. The central governing body of the Church, the consistory, was composed of ministers and city councilors co-opted by the clergy. Another body, the venerable company of pastors, was composed of all the pastors and twelve elders selected by the councils. This body proposed legislation to the consistory and also oversaw the appointment of pastors.[58]

Church government and discipline was to be applied territorially. The church building was to be kept, but not consecrated. Calvin had the same ambivalence to the structure as did Luther. He believed churches were necessary because of the need to provide for a community assembly but they were not especially holy and they were not to be called houses of God: 'Let those who tie down the church to power in its ordinary sense, and to pomp, hear what Hilory says on that subject, "we do wrong in venerating the Church of God in Roofs and Edifices."'[59] Churches were simply places of worship and because of the tight strictures on public behavior, the actual range of permissible conduct within them was narrow. Rural churches were to be locked when not used for worship.[60]

Calvin, for the same reasons as Luther, retained the parish as a means of defining and molding the community. Before the Reformation, Geneva was divided into five parishes. Calvin redrew the boundaries to divide the city into three. Supervision of the community by the consistory was to occur in part territorially. Two members of the consistory, accompanied by the pastor, were to visit each parish regularly so that 'their eyes might be on the people.'[61] The parish boundary was essential for assigning critical functions; 'for bringing children to catechism and for receiving the sacrament, the boundaries of the parish as far as possible should be obeyed.'[62]

Other Protestant reformers went further than Calvin and Luther not only in attacking Church organizations but in attempting to avoid creating new ones; a few even attacked the parish. John Smythe for example in his 'Principles and Inferences Concerning the Visible Church', 1607, sought to define the true manifestations of a Christian Church government. A visible communion of the genuinely faithful or 'saints,' Smythe said, would produce a visible Church. 'The visible church is the only religious society that God hath ordeyned for men on earth. All religious societies except that of a visible church are unlawful [and among the more notorious of these are]: Abbeyes, monasteries, Nunries, Cathedrals, Collegiats, [and] *parishes*.'[63]

The earliest Baptist and Congregationalist elements were not organized along parish lines and their views on church building were far more flexible than the Lutherans and Calvinists. Perhaps the Friends or Quakers were and remain the group with the least hierarchy; and, not surprisingly, their divisions, even in Europe, were not based on parishes nor were their buildings consecrated or specified solely for worship. They were 'meeting' houses. Yet even among the Quakers there tended to be some slight spatial or territorial reflection of hierarchy. Meeting houses often had seats set apart for the elders of the group.

The unusual conditions within the United States of the separation of Church and state has spread the organizational continuum of the Christian denominations even farther along the direction of less hierarchy and rigidity and less territoriality.[64] With the exception of Episcopalians and some Methodist denominations, twentieth-century American Protestant faiths do not refer to parishes or dioceses in describing their structures.[65] Congregants are conceived of in non-territorial, though often spatial, terms. Congregations can form 'nodal' regions. A church may decide not to locate near another of the same denomination for fear there may not be enough congregants to fill both. But which building a person will select to attend is not a matter of domicile, but rather of convenience.

Whereas the basis of attendance is not linked to rules about residence, some Protestant Churches do use territories to define domains of responsibility among their higher officials. In the weakest sense territories can be used simply as devices for grouping churches and congregations when it comes to sending delegates to higher Church councils. Such mild territoriality is used by Quakers in selecting delegates to national meetings. A bit stronger are those few Protestant denominations which have someone like a bishop overseeing the conduct of ministers within a territorial unit.

Post-Reformation Church

At the extreme end of the territorial continuum still lies the Catholic Church. But even here there has recently been a relaxation in territorial control at the level of parishioners. It is now possible for Catholics to attend mass at churches other than that of their parish. It is also possible to have marriages, baptisms, and communions in a church other than one's own if permission is received from the proper Church officials. These conditions were initiated to reflect a more geographically mobile population. Still, the Church hierarchy remains as territorial as before. Indeed, for the last 400 years overall Catholic Church organization compared to other Churches has appeared less modern or more traditional.[66]

The Reformation was part of radical historical changes; among them the rise of capitalism, of individualism, and the questioning of traditional authority. These were powerful developments. Yet, unlike the social changes in the Classical and Feudal Worlds which had profound effects on the Church, these modern developments, by and large, had relatively few repercussions on Catholic Church doctrine and structure. In its efforts to regain the ground it lost in the Reformation, and in its subsequent expansion in the New World, Africa, and Asia, the Church continued to consolidate its authority as well as fight old battles such as issues of enfeudation. It took steps towards establishing greater impersonality in Church discipline and office, and stricter enforcement of appointment by examination, meritorious promotion, and standardization of payment by wage or stipend. Despite the fact that these became the elements modern governments and business incorporated into their own bureaucracies, the Church's vast and conservative economic holdings, its commitment to other organizational forms such as the model of the family, its commitment to tradition and ritual, and its loss of influence in those areas of the world that were at the vanguard of economic development prevented Church structure from developing further along modern lines and prevented the Church from becoming a force for change.

We have now seen how and why the Church employed territoriality and we have seen how its effects were integral components of Church organization and helped to form those facets of Church structure that appear modern, especially those concerning impersonal hierarchical relationships. But the effects of impersonality within Church hierarchy could progress only so far before coming into conflict with the Church's commitment to tradition, ritual, and other organizational models. The two other major contributions of territoriality to modernity – obscuring sources of power, and presenting an emptiable and fillable space – were of even less use to the Church. The Church was far more interested in making its power visible than in disguising it. This meant relying on reification and displacement which, as was noted, is not the same as obscuring sources of power.

Repeated emptying and filling of space also was not to be an important use of Church territory. At the large geographical scale this use comes about when sufficient political power exists to actually plan, move, settle, and remove large populations. Although the Church did have the backing of Spain and Portugal to redistribute populations of Native Americans and Asians, these Church-directed efforts were small and short lived compared with those linked to the political territorial ventures of the colonial powers.[67]

The Church did possess power enough to control what occurred within its buildings. But at this scale it was traditional architectural form and ritual that prevented the church space from being conceived of as a conceptually emptiable and fillable mold. The church building was indeed often without parishioners. But this did not mean that it was then thought of as a vacuum, potentially fillable with unspecified activities. Rather every detail of the church building contained symbolic meaning and pre-assigned uses (though secular activities took place within it as we saw in the Middle Ages), and the entire structure, as a consecrated place, was never without spiritual content. This again is quite different from the modern architectural sense that buildings can be designed as shells containing volumes or spaces which can be put to multiple and interchangeable uses. Indeed, to have the territorial power to empty and fill space and conceive of space as a contingent framework inhibits the sense that space and place are filled with experience and spiritual import.

The territorial effects of obscuring sources of conflict and of conceptually and actually emptying and filling space, as well as of creating impersonal relationships, are the most important modern components of territoriality. To explore them we will turn to realms that exercised enormous power over spatial behavior while de-emphasizing the content and meaning of place and space: the realms of politics and work. Political organization and work were the areas most intimately associated with changes in capitalism and each exhibited modern territorial effects but at different geographical scales.

5
The American territorial system

The Reformation was but one upheaval occasioned by the Renaissance and the rise of capitalism. Others were world exploration, discovery, and colonizations. The vast North American hemisphere, with its relatively sparse indigenous population at the time of contact, provided Protestant Northern European colonials with the most fertile grounds for developing a modern political economy. How and why did territoriality become an instrument of settlement and government and what are its modern components?

Discovery and colonization

Emergence of abstract territory and space

Modern uses of territory did not appear all at once full-blown. The original grants and charters for the New World were primarily feudal documents in which monarchs bestowed both political and economic privileges to persons and companies.[1] Yet the conditions the New World presented were so different from previous cases of territorial expansion that the old formulas were forced to have new twists. From the territorial perspective, what stands out so starkly to modern eyes about the very beginning of the 'Discoveries' is the abstract geometrical nature to the claims of sovereignty over area. These claims appear to be the natural precondition for clearing a place for community and authority and for molding further and more specific social organizations. In its scale and intensity, no less than in its conception, this approach to people and place has a modern ring. It points to an explicit and intense territorial definition of social relations. When the relationships between people and territory change frequently the territorial definition leads to an abstract sense of territory and of space, one which is conceptually separable from the events it contains. In short, from the very beginning, Europeans appeared to be employing territoriality to a significant degree

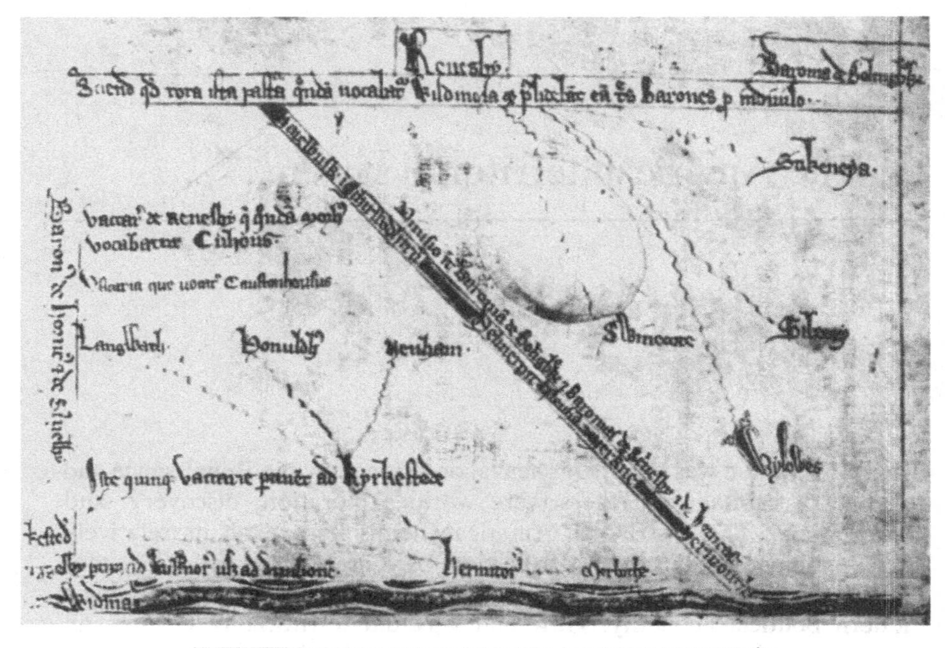

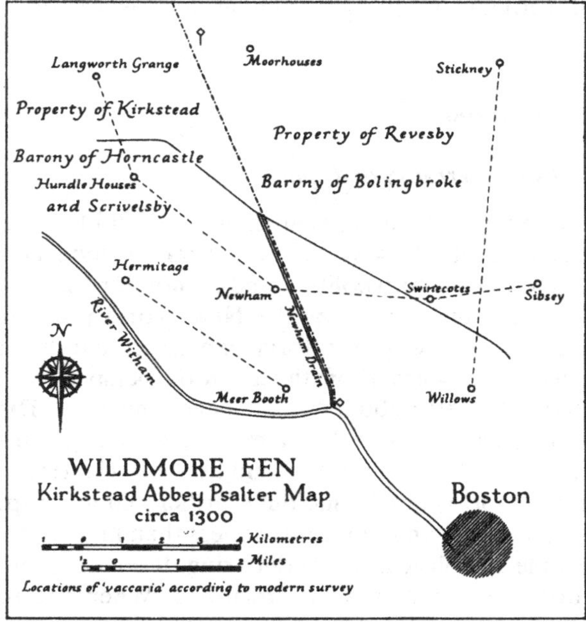

Figure 5.1 Medieval cadastral maps

(a) Plan of the boundary between the manors of Kirkstead and Revesby. From an account c. 1300 of a survey made c. 1150. From the Kirkstead Psalter, now at Beaumont College, Old Windsor.

(b) The same boundary and 'vaccaria' as indicated on the modern survey abstractly and this both presupposes and reinforces an abstract conception of space.

The presence in the sixteenth century of a more modern use of territory and space is clear in retrospect, yet the contemporary conceptions were in fact a complex blend of old and new. Most observers were barely aware that changes were taking place. Indeed, the characteristics of space and territory we associate with modern uses are to this day still unfolding and have penetrated different levels of social consciousness to different degrees and at different periods. These finer points about stages of development and levels of penetration will be addressed but we must first consider the overall initial appearance of the modern view of space.

The discoveries, and the newer meanings of space and territory, were part of two fundamental social changes: the replacement of the old economic order of feudalism with capitalism; and the replacement of the medieval mentality by the Renaissance. Merchant capitalism led Europe away from a fragmented, cellular, feudal economy and polity to a global economic network based on a handful of national political and economic systems seeking new markets, new transportation routes, and enlarged territorial jurisdictions. And from a traditional and other worldly intellectual orientation emerged a vital curiosity about the world, an insatiable appetite for new experiences – and Protestantism to justify them.

The view of earth space that predominated in the thousand years between the fall of Rome and the Age of Discovery was derived from the geographic experience of a closed feudal society and from the influence of the Catholic Church's interpretation of the Bible. The former provided an ordinary conception of space, place, and distance closely linked to the everyday experience of their contents. It was based on the same sense of space that we described for primitive and peasant societies. Events and experiences, not location in abstract space, defined places and distances. Though it was possible to measure roughly the distances between places, these abstract units had little relationship to concrete experience. Moreover, the Medieval World simply did not have the technical apparatus to represent accurately even local areas. Medieval maps of land holdings were sketches (see Figure 5.1) containing numerous distortions and scales.[2] When distant places were collected together within a conceptually abstract system it was in conformity with a religious cosmography. The visual representation of this world view usually had the earth divided into three regions: Africa, Asia, and Europe;

maps. Guide lines show how the Psalter map gives with reasonable accuracy the positions of the vaccaria relative to each other and relative also to the 'straightened' River Witham

Source: 'Medieval Surveying and Maps,' *Geographical Journal*, 121 (1955). Reproduced by permission of The Royal Geographical Society, London.

and these were often centered on Jerusalem[3] (see Figure 3.8). The import of this conception was that wherever one actually is located in physical space is irrelevant; one should strive to be in the center – at Jerusalem.

Instead of using this closed symbolic system, the explorers were thinking of earth space (and time) more abstractly and geometrically: longitude, latitude, distance, and time were the primary spatial referents.[4] Examples can be found in the West and in other civilizations of people using coordinate systems to describe the earth. But as we noted these had a different import and were used for different purposes than was the case for the emerging Western system of spatial relations. The Chinese coordinate system was infused with cosmic meanings; each square was a microcosm of the world.[5] The grids used to map out plans of towns and land holdings in the Roman Empire (centuriation) did not suggest a conception of an abstract spatial continuum. Rather, the units were bounded and were often linked to pre-existing land holdings and conformed to local topography.[6]

Time as well as space was slowly altered by post-Renaissance events. No longer was the world simply passing time until the second coming of Christ. The rise and fall of the Ancient World had been rediscovered. Europeans were now aware that civilizations change through time. Indeed, the new economic order was about to propel the West into an era of rapid and unparalleled change. The present society was coming to be understood as different from past societies and the future would have even greater differences in store. For many, these changes were thought to be for the best; the passage of time was about to become the road to progress. As a measure of change and progress, time became abstracted from particular events and this abstraction increased as time was to become more finely measured and units of work were to become defined by it.

These changes in awareness of earth space and time did not have an immediate effect on Western political and theoretical speculation about territory and society. Renaissance thought about territory continued in the medieval fashion to be limited to two principal areas. One was the relationship of geographic size to political process. Another was the merits of different forms of land ownership or property rights. These were discussed abstractly, but the general understanding was that despite their origins, territorial relationships were primarily socially defined. They tended to circumscribe real 'organic' communities. A community in general was thought to be a unified entity with a common purpose. Whereas many non-geographical communities existed (the community of Christians for one), territories were generally believed to bind cohesive social entities. For feudalism this was the case especially at the lowest geographical scale – the estate and the town – where the political and economic were coterminable. Here the community and the territory were practically synonymous. This association is realized clearly in the prevailing contemporary views on community representation.

Since the essential needs and goals of the community were believed to be shared by its members, adequate representation would result if someone could convey these needs to the government. That someone, whether elected or not, was a representative. This system was the customary form before the seventeenth century but does not seem to have ever received a name. We will call it 'organic representation' although it does possess characteristics similar to those described by Burke as 'virtual representation.'[7] Unlike the modern view of proportional representation in which a representative stands for a specified number of people, organic representation did not see the task of representation to reflect the numbers of individuals in the community. Each community was thought to be a single entity with a common will and need, and therefore each community, regardless of its population, could be represented by even a single person who was knowledgeable about and sensitive to its interests. That person need not even be from the community he represented. It was not until the eighteenth century (as demonstrated in local constitutions adopting proportional representation) that such a view was seriously challenged, though not entirely replaced. By then it became overwhelmingly evident that geographical communities were no longer necessarily united in their interests but rather were often transient collections of individuals with different and competing interests and needs. With this more territorial definition of social relations came the need to conceive of a system with some sort of proportional representation.

The Old World viewed territoriality primarily as socially defined, but events were about to change this. Awareness of the New World accelerated an abstraction of space because the Americas presented European powers with a vast, distant, unknown, and novel space. This meant that with the limited technology and political power at their disposal, Europeans could still claim to 'clear' the space and form territories to organize and fill it at all geographical levels and with an intensity that was impossible to match in the Old World. Again it is important to note this realization and use of space did not occur at once. It is in fact still emerging and intensifying. The discoveries, however, gave the process an enormous boost.

Colonial claims

The 'Alexandrine' Papal Bulls of 1493 are the first examples of what appear to us as, and what was over a number of generations to become, in effect, a stark abstract metrical territorial definition of social relationships. These documents, soon supplanted by the Treaty of Tordesillas (1493), granted to

Spain authority over any newly discovered non-Christian lands approximately 100 leagues to the west of the Azores and the Cape Verde Islands, and to Portugal authority over new discoveries to the east.[8] This line was in effect a latitude drawn from pole to pole. For the first time in history an abstract geometric system had been used to define a vast – global – area of control. To the fifteenth century this may not have appeared so grand and abstract a geometric division of the world because the discoveries of Columbus and his immediate successors were not thought to be continents but rather numerous islands off the coast of Asia. The term island appears frequently in the Treaty of Tordesillas itself and it is through islands that the Pope justified his involvement because, according to the Church, the 'spurious' eighth-century Donation of Constantine included a bequest to the Church of islands of the western portion of the Roman Empire. By extending the boundary of the old Empire to include the Atlantic Ocean, this document was interpreted by the Popes to mean that they had authority to grant dominion over islands in that sea as well as in the Mediterranean.[9]

The Papal grants of dominion to Spain and Portugal were encumbered with both religious and secular obligations. On the religious side it was expected that Spain and Portugal would have authority and responsibility for all of the souls within their respective dominions. As a secular grant it was interpreted as an extension of feudal authority over whatever existing political entities these territories might contain.

Not only was the vastness of the New World underestimated, but also the earliest charters and grants indicate that the late fifteenth to the middle sixteenth centuries expected the inhabitants of these realms to possess customs and territorial units that were familiar to those of Europeans.[10] Conquering would then require establishing suzerainty over these people and their lands. The supposition that the new lands would contain Old World types of political organizations is seen in the languages of the early charters. The Letters Patent of King Henry VII to John Cabot, 1496, declare that Cabot may 'subdue, occupy and possesse all such townes, cities, castles and isles of them found which they can subdue, occupy, and possesse as our vassals, and lieutenants, getting unto us the rule, title, and jurisdiction of the same villages, townes and castles and from land so found,'[11] and the Charter to Sir Walter Raleigh, 1584, gave him authority over lands, territories, and countries that were conquered and over their 'cities, castles, townes and villages.'[12]

Although explorers came in contact with high civilizations in Central and South America, they did not find such 'cities, castles, townes, and villages' in the form expected in much of North America, and gradually the way was opened for the North American territories to become presented to the public as a wilderness, and, in terms of their social content, as 'empty space.' Many of the native North Americans that Europeans came upon had 'primitive' economies with communal land use. They practiced agriculture but also were hunters, gatherers, and collectors. Their fields appeared vague and untidy to European eyes and, with the exception of the dog, they did not have domesticated animals. Their customs and land uses seemed so alien and their political processes so inconspicuous that many Europeans concluded that Indians were sub-human and could and ought to be removed from the land.

In his 'Generall Historie of Virginia, New England, and The Summer Isles,' John Smith describes the Indians (in an about face, for he had written before that they were gentle and generous) as 'perfidious and unhumane people; cruell beasts [with] a more unnaturall brutishness than beasts,'[13] and Smith's friend, Purchase, argued (according to Jennings) that 'Christian Englishmen might rightfully seize Indian lands because God had intended his land to be cultivated and not be left in the condition of "that unmanned wild Countrey, which they [the savages] range rather than inhabite." '[14] Although this idea of 'range rather than inhabite' was contrary to fact, it was to become a rationalization for European expropriations of Indian lands (and was to be used throughout European colonization as an excuse to remove aboriginal peoples everywhere). Defining Indians as wild or sub-human was an important step in their removal because contemporary Europeans generally held that 'to invade and dispossess the people of an unoffending civilized country would violate morality and transgress the principles of international law.'[15] Conquered Indians were treated differently than were Europeans. Most of the Dutch settlers in New York were not dispossessed of their lands when the English took over the colony.[16] Rather they were required simply to obey English rather than Dutch law. Not all the Irish were removed from their lands when the English planted colonies in Ulster.[17] The Dutch and Irish were Christian and the English thought the former, especially, to be a civilized people whose property rights deserved respect. Believing that the Indians were uncivilized and savage, more than anything else, diminished Europeans' responsiveness to Indian claims to land and made the great expanse of the Americas appear to be empty space for the taking.[18]

Of course the expungement of Indians was easier thought than done. Indians, far outnumbering the first European settlers, were often able to fend them off. In many cases the two groups lived intermixed, sometimes in friendship, more often not, and often affecting each other's cultures. Some Europeans spoke out against the appropriation of Indian lands. But these were usually self-serving arguments and those making them would be happy to possess Indian lands if the opportunity arose. Roger Williams, who tried to protect Salem's right to lands that it purchased from the Indians rather than from the English Crown, had to justifiy the fact that the land was owned by the Indians. According to Cronon, Williams argued that the King had committed an 'injustice, in giving the Countrey to his *English* Subjects, which belonged to the Native *Indians*.' Even if the Indians used their land differently than did the English they nevertheless possessed it by right of first occupancy and by right of the ecological changes they had wrought in it. Whether or not the Indians conducted agriculture, they 'hunted all the Countrey over, and for the expidition of their hunting voyages, they burnt up all the underwoods in the Countrey, once or twice a yeare.' Burning the woods, according to Williams, was an improvement that gave the Indians as much right to the soil as the King of England could claim to the royal forests. If the English could invade Indian hunting grounds and claim right of ownership over them because they were unimproved, then the Indians could do likewise in the royal game parks.[19]

But arguments for Indian rights, whether self-serving or not, were drowned out by the clamor for Indian land. By the beginning of the seventeenth century the characterization of Indians and their non-agricultural relationships to land as sub-human served as the rationale for white expansion. In 1629, Governor Winthrop of Massachusetts declared that most land in America 'fell under the legal rubric of *vacuum domicilium* because the Indians had not "subdued" it,'[20] and John Cotton wrote that 'In a vacant soyle, he that taketh possession of it, and bestoweth culture and husbandry upon it, his Right it is.'[21] This position persisted through the nineteenth century. Chief Justice Marshall in 1823 argued that the Indians 'were fierce savages . . . whose occupation was war, and whose subsistence was drawn chiefly from the forest . . . [and therefore] that law which regulates . . . the relations between the conqueror and the conquered was incapable of application to a people under such circumstances.'[22]

The New World charters described territorial claims abstractly and geometrically and, in conjunction with conceptually and then actually clearing the land of Indians, the geometric lines of territorial authority become sweeping space-clearing and maintaining devices for territorially instituting communities. The first Charter of Virginia (1606) established jurisdiction over a territory carved out by lines of latitude, between 34° and 45° North and up to 100 miles off shore. It called for establishing two settlements somewhere in this metrical space, one between 34° and 41° North, and the other between 38° and 45° North.[23] Each had a claim to the resources 50 miles North and South and 100 miles West. These claims were elaborated and extended in the second (1609) and third (1611–12) Charters. The third extended offshore claims to 300 leagues, but neither contained details about how the territories were to be further subdivided.

On February 3, 1618, the company gave the colony authority to grant incorporations and in 1619 the proceedings of the first Virginia Assembly mention that representatives attended from each incorporation and plantation. On the 24th of August, 1621, the Virginia ordinances established a General Assembly based on representation by two burgesses from every town, hundred, or other 'particular plantation.' Thus by as early as 1618, and certainly by 1621, a colony containing only a few thousand people established territorial subdivisions for political representation. By 1634, the General Assembly created the county form of local government and thus a full-fledged hierarchical territorial organization was in place.[24] The Virginia Charters used abstract metrical boundaries but did not elaborate on the types of territorial subdivisions. The first charters of subsequent colonies were to do both and on occasion described procedures for areal representation. Indeed, they appeared often to take on flexible and even experimental approaches to territorially subdividing space.

The Charter of 1629–30 to Sir Robert Heath defined his domain as

> all that territory or track of ground, scituate, lying and being within or dominions of America, extending from the north end of the island called Lucke island, which lieth on the southern Virginia seas, and within six and thirty degrees of the northern latitude, and to the west as far as the south seas, and so southerly as far as the river St. Matthias, which bordereth upon the coast of Florida, and within one and thirty degrees of northern latitude, and so west in a direct line as far as the south sea aforesaid.

It gave to him full and free power to 'erect villages into boroughs, and boroughs into cities for the merits of the inhabitants and the convenience of the places with privileges and befitting immunities to be erected and incorporated, and to do all other and singular, upon the premises which shall seem most convenient to him or them.'[25] 'A Declaration and Proposal of the Land Proprietor of Carolina,' 1663, article 4, contains stipulations for sub-territorial units and for areal representation. It states that

> We shall . . . empower the major part of the freeholders or their deputies or assembly-man, to be by them chosen out of themselves, viz; two out of every tribe, division, or parish, in such manner as shall be agreed on, to make their own laws, by and with the advise and consent of the governor and council.[26]

The Charter of Rhode Island, 1663 (its boundaries set out in previous charters), charges the General Assembly with erecting and setting up places and courts of jurisdictions and 'prescribing, limiting, and distinguishing the number and bounds of all places, townes, or cityes.'[27] The Grant of the Province of Maine, 1639, allows for the incorporation of cities, boroughs, and towns with all liberties and things belonging to the same

> [and] grants full power and authoritie to divide all or anie part of the Territories hereby granted . . . into Provinces, Counties, Citties, Towns, Hundreds and Parishes or such other partes or porcons as hee . . . [sees] fitt and in . . . every or any of them to appoynt and allott out such porcons of . . . land for publique uses.[28]

The Charter for New Caesarea or New Jersey, 1664, created a General Assembly to lay equal taxes . . . within the several precincts, hundreds, parishes, manors, or whatever other divisions shall hereafter be made and established in the said Province . . . [and] to divide the said province into hundreds, parishes, tribes, or such other divisions, and districtions as they shall think fit, and that representation should be based on these areal divisions.[29]

In the Fundamental Constitution of the Province of East New Jersey, 1683, is found even more flexible geographical approaches to representation. Representation to the great council is outlined but without specifics because the territorial units are not yet fixed in number and kind:

forasmuch as there are not at present so many towns built as there may be hereafter, nor the Province divided into such counties as it may be hereafter divided into, and that consequently no certain devision can be made how many shall be chosen for each town and county; [though] at present four and twenty shall be chosen for the eight towns . . . and eight and forty for the county.[30]

The Charter to William Penn, 1681, is remarkably metrical. It gave to him all the land

bounded on the East by the Delaware River, from twelve miles distance Northwards of New Castle Towne unto the *three and fortieth degree of Northern Latitude*, and from the head of the said River, the Easterne Bounds are to bee determined by a *Meridian Line*, to bee drawne from the head of the said River, unto the said *three and fortieth degree*. The said Lands to extend westwards *five degrees in longitude*, to bee computed from the said Easterne Bounds; and the said Lands to bee bounded on the North by the beginning of the *three and fortieth degree of Northern Latitude*, and on the South by a Circle drawne at twelve miles distance from New Castle Northward and Westward unto the beginning of the *fortieth degree of Northern Latitude*, and then by a *streight Line Westward to the Limitt of Longitude* above mentioned.

Within this domain Penn was granted the absolute power 'to Divide the said Country . . . into Townes, Hundreds, and Counties, and to erect and incorporate Townes into Borroughs and Borroughs into Cities.'[31] Penn attempted to do this in his elaborate 'Frames of Government' (1682–96 and 1701) which contained detailed schemes for sub-territorial units.

The 1669 plans for Carolina were perhaps the most fantastic and elaborate scheme for hierarchical territorial partitioning. It was supposed to have been drafted by Locke but was never fully put into effect. The extent to which hierarchy is stipulated can be seen from the first of many rules regarding land subdivisions:

The whole province shall be divided into counties; [forming squares] each county shall consist of eight signiories, eight baronies, and four precincts; [and] each precinct shall consist of six colonies . . . Each signiory, barony, and colony shall consist of twelve thousand acres . . . so that in setting out and planting the lands, the balance of the government may be preserved.[32]

These charters and grants offer evidence for abstract and even experimental attitudes concerning territory and community at a large geographical scale. What can be said about attitudes towards community and land at a more local level? Though the answer is complex, here too we find evidence for a progressively more abstract conception of space. Although the early grants contained several feudal characteristics, the type of land tenure they most often stipulated was modeled after the tenure of East Greenwich in the county of Kent. (The grants of land in Maryland and Pennsylvania were modeled after the tenure in the Bishopric of Durham which was much like the Kentish.) The general characteristics of Kentish gavelkind, or free and common socage tenures as they were called, were that:

> 1. the land was freely partible in equal shares among the male heirs since there was no primogeniture in the descent of property in Kent. 2. Kentishmen could freely sell or give their lands, and could sue for the same in the King's court, even against their lords. 3. Court procedures were more direct and less cumbersome than elsewhere and 4. All Kentishmen were born free; villeinage was unknown in Kent.[33]

Granting land as of the practice in Kent meant that the least feudally encumbered system of land tenure would be practiced in the North American colonies. This type of land ownership was the one best suited for speculative commercial ventures involving a mobile population quickly buying and selling unfamiliar, unexplored land, and for conceiving of it as quantifiable and valuable parcels of space.

Whereas this unencumbered land system encouraged an abstract conception of place, the first settlers had to come to know the land and wanted to put down roots. But their geographical knowledge was limited, especially so because there were few of them and the land was vast. Some degree of acquaintance is shown in the early land claims for counties, townships, and private land holdings. Land claims often mentioned natural features such as a coastline, a river, a marsh, or a hill. Frequently Indian place names and descriptions were included in early European land claims. For example the 1636 deed of land bought from the Indians by William Pychon describes the property at Springfield as

> that ground and muckeosquittaj or madows, accomsick, viz: on the other side of Quana; and al the ground and muckeosquittaj on the side of Agaam, except Cottinackeesh or ground that is now planted . . . [and] al that ground on the East side of Quinnecticot River . . . reaching about four or five miles in Length from the north end of Masaksicke up to Chickuppe river.[34]

An individual would come to know the contents of his farm land and some of his own township, but the vastness of the land and the sparseness of its European inhabitants make it unlikely that he would know well the adjacent area out of which might come the next town. Indeed, the formation of towns was often irregular, leaving open, unclaimed spaces. It was not until the early eighteenth century that Massachusetts required settlements to be contiguous for greater ease in protecting settlers from Indian uprisings.[35] And even if it were the area directly adjacent that was to be settled, it still did not mean that those who were to settle it knew it first hand. Land was often disposed of before it was surveyed.[36] It was as late as 1715 that Massachusetts made its first concerted effort to survey lands before they were disposed of, in this case lands between the Connecticut and Merrimac Rivers.[37] Surveys at this period described land in the barest of terms, in part because they were becoming abstract and geometrical.[38] The increased spatial precision gained in the mid-eighteenth century by describing land in terms of rectangulation or simple distances and directions from a single point was often acquired at the expense of mentioning its local physical features and history. For instance, the type of physical and cultural detail (though modest) contained in the 1636 deed to William Pychon (above) is absent in an eighteenth-century deed from the same county which transferred rights to two entire townships defined abstractly as

> the full Contents of Six miles in Width and Seven miles in Length and . . . Townships to be laid out in a regular form the Southwesterly Corner of Said Tract of Land beginning at ye North Easterly Corner of a New Township called and known by ye name of New Framingham and from thence running so far Northerly and Easterly as to include both ye above described Townships and lay them out in a Regular form as abovesaid.[39]

North America and Ireland

A brief comparison of the processes in North America with those accompanying the English colonization of Ireland (which was the other major area of English overseas expansion at the time) illustrates how much more abstract were the North American territorial claims, and how much more empty the land appeared. The English had invaded Ireland in the twelfth century and ever since made claims of one type or another over Irish land. Knowledge about Irish customs and geography was available to the English and although some Irish were not agriculturalists they were all Christians, though to the Protestant English after the reign of Henry VIII perhaps the wrong kind. Indeed they had been Christians longer than the English.

England's long acquaintance and contact with Ireland, and her begrudging acceptance of the Irish as humans and Christians, added specific and often minute detail and meaning to the English image of Irish landscape that was impossible conceptually to erase. That this familiarity made it difficult for the English to think of Ireland as an empty space which could be neatly geometrically subdivided into territorial units is seen in the degree to which maps, surveys, and plans of English plantations in Ireland conformed to the rich pre-existing Irish context.[40]

The English surveyed Irish lands before they undertook major efforts at colonization in the late sixteenth and seventeenth centuries. These, as well as records of deeds of property, point to the degree to which the English plantations were grafted onto pre-existing Irish land holdings.[41] Plans often included the interspersion of Irish within the English plantations. For example, the English decree preceding the 1608 orders laying the foundation for the plantation of escheated lands in Ulster states in part four that 'the undertakers of these lands be of several sorts. 1. English and Scottish who are to plant their portions with English and Scottish Tenants. 2. Serviters in Ireland who may take English or Irish tenants at their choice and 3. Natives of these counties, who are to be free holders.'[42]

Then follows a county by county survey of land. To prevent the need for future resurveying, there was contained the stipulation that pre-existing land measures and units should be used: In the words of the decree 'To avoid confusion . . . and further charge of measuring the whole county, every ballyboe [Irish unit of area] is to have the same bounds and quantity, as were known, set out, and used at the time of the departure of the late Traitor Tyrone.'[43] Further instructions about conforming to pre-existing land uses are found in the 'Orders and Conditions to be Observed by the Undertakers Upon the Distribution and Plantation of the Escheated Lands in Ulster' (from a copy printed in 1608), and in 'Articles for Instructions to Such as Shall be Appointed Commissioners for the Plantation of Ireland.' These, for example, express the need to be careful to distinguish escheated lands from other lands, and that land division within the counties should recognize the 'ancient limits of the old parishes' or be 'bounded out by the known Metts and Names, with the particular Mention both of the Number and Name of every Ballyboe, Tath, Polle, Quarter or the like Irish precinct of land, and to give each Portion a proper Name to be known by.'[44]

The English surveys of Ulster in 1618 (ten years after the plans for settlement) further illustrate how very detailed was the knowledge of the land and how much of the native population and knowledge of the past was incorporated into the new settlements. A typical example is in Pynnar's survey. For the county of Cavan, Pynnar begins with mention of the 3,000 acres held by Sir James Hamilton. The names of the principal lands are Keneth, 2,000 acres, and Casbell, 1,000 acres. Pynnar describes the Casbell land as follows:

> Upon this Proportion is built a very large strong Castle of Lyme and Stone, called Castle Aubignie, with the King's Arms cut in Free-Stone over the Gate. This Castle is five stories high, with four round Towers for Flankers, the body of the Castle fifty feet long, and twenty eight feet broad . . . There is adjoining to the one End of the Castle a Bawne of Lyme and Stone eighty Feet square, with two Flankers fifteen Feet

high. This is very strongly built, and surely wrought. In this castle himself dwelleth, and keepeth house with his Lady and Family. This Castle standeth upon a meeting of five beaten ways, which keeps all that Part of the Country.[45]

He goes on to describe the number of British freeholders and cottagers on this land, their land holdings and characteristics, as well as to survey the lands allotted to servitors and natives.[46]

English settlers in Ireland could know much more than did those in North America about the land in which they were settling. Maps, surveys, and deeds continued to incorporate detailed landscape features and pre-existing Irish land use practices and divisions.[47] Moreover, the settlement plans had Irish populations interspersed among the English. Compare this with the abstract claims of territorial control made by the charters and grants of North America. Here a physical feature may be mentioned, there an Indian place name, but the overwhelming sense is of claiming and subdividing an empty space.

Diminished social definitions of territory

At their founding, then, practically every colony in British North America was using abstract geometrical forms to claim land, and was preparing the way to subdivide its territory into convenient levels and numbers of sub-territorial units. That even the sub-territories were thought by contemporaries to be at least in part defining social relations is seen in the frequency with which the word 'convenience' is used to justify their establishment. What is more, the freedom given to the colonists to determine the most convenient type, level, and number of sub-territorial units suggests also that the relationships between people and territory is in some respects being posed as a gigantic social experiment in the use of space, to affect, organize, and control behavior.[48]

Yet accompanying the territorial creation of social units was the feeling that these units ought to encompass organic social entities. But the vitality of this view was to diminish as the role of commerce and geographical mobility increased. The losing struggle to retain a socially defined sense of community at the local level is illustrated all over the North American continent but perhaps most sharply by the Puritan case.

Puritans settled in New England to establish theocracies, and the parish played a large part in defining and organizing community. The New England town was a parish with civil authority grafted on. Yet the dynamic character of events in the New World were soon recognized by the settlers and were incorporated into Church structure. By 1648, Puritans established rules by which preachers and parishioners could change parishes.[49] But these concessions to mobility were not enough always to keep community and parish a single entity. Parish boundaries, especially in cities, were often ignored.

Splinter churches arose which did not coincide with pre-existing territories, and some denominations did not recognize parishes at all.[50]

Other indices of a loosening of a social definition of territory can be seen in the increasing emphasis on residency as a definition of membership in New England communities. Among the reasons for keeping New England society closed were the desire that they contain citizens of compatible religious views and that they not be heavily encumbered as they were in England with the burdens of supporting the poor. The most direct means of keeping out those who were undesirable was to create stringent legal requirements for residency. Town lands belonged initially to the founding settlers, and newcomers had to be approved by the townsmen before they could settle in the community and hold land. This meant that the town would care for only those poor who were proper members of the community.

One of the early recorded attempts at restricting entry into a New England town is found in a 1639 rule for the village of Sandwich which established a general principle of entry as follows: 'And for the better carrying on of afaires among them, it is therefore ordered, that noe man shall hence forth bee admitted an inhabitant into Sandwich, or injoy the privileges thereof, without the approbation of the Church, and Mr. Theo Prence, or any of the Assistants whoe they shall choose.'[51] Similar rules were enacted for other towns but it was often the case that if another town would guarantee to receive back its emigré, then he could remain for a while; but sooner or later he had to apply for admission to the town in which he now resided.[52] As early as 1636, Plymouth Colony established rules for the newcomer, stipulating that 'noe person coming from other ptes bee alowed an Inhabitant of this jurisdiction but by the approbacon of the govr and two of the magistrates att least,'[53] and in the same year Boston enacted regulations to control even the stay of guests ordering that 'noe Townsmen shall entertaine any strangers into theire houses for above 14 dayes, without leave from those that are appointed to order the townes businesses.'[54]

In order further to protect entry, many communities stipulated that no one was allowed to sell land to strangers without community approval. In a dispute over land transactions, the town of Sudbury, in 1636, contended that Richard Fairbanke 'hath sould unto twoe strangers the twoe houses in Sudbury end that were William Balstones, contrary to a former order, and therefore the sayle to bee void, and the said Richard Fairbanke to forfeite for his breaking thereof.'[55] And in 1657 Plymouth Court responding to the complaints of the town of Taunton to the effect that

> sundry unworthy and defamed psons have thrust themselves into the said towne to inhabit there, not haveing approbacon of any two magistrates according to an order of the court, and contrary to the minds of the divers of the inhabitants, . . . [determined] 1. that noe such pson bee intertained by any inhabitant of the towne, or the penaltie [be levied] of forfeiting twenty shillinges for every weeke that they shall

intertaine them without the approbacon of the five selectmen appointed to order the publicke affaires of the towne: . . . [and] 2. Likewise it is ordered, that you give warning to youer townsmen, that noe pson or psons of youer towne do sell, hier, or give house or land to any pson, so as thereby to bring them in to bee inhabitants amongst them, but such as have approbacon of two of the magistrates att least, according to an ancient order of court, as they will answare theire contempt in doeing the contrary.[56]

Many communities adopted such restrictions on membership in order to control access to ownership of land and especially to control eligibility for poor relief. But most communities were becoming larger, and strangers entered them in ever increasing numbers. To keep order and provide social services, a more lenient criterion for residency was needed. A consensus formed based on a clause in a law enacted in Plymouth, 1642, stating that 'every pson that liveth and is quietly settled in any township and not excepted against within the compass of three months after his comeing in this case shall be reputed an inhabitent of that place.'[57] 'Warning out', as it was called, had the effect of shifting to territoriality some of the definition of community membership. Simply being in the place for three months without being found out or without making oneself a nuisance, allowed one to become a member of the community and eligible for poor relief. Warning out was soon adopted by other New England communities and was incorporated within the 1672 Articles of Confederation (Article XIII). But population and geographical mobility continued to increase to the point where some communities simply could not afford to support all of the poor who entered and who were not warned out. So the eighteenth century saw a lengthening of time a person must reside before being accepted into the community and being eligible for poor relief. Lengthening the time was also calculated to make the colonial, and later the state, governments bear part of the burden of poor relief.[58] The states came to do just that in the nineteenth century and this permitted local communities to shorten the period for residency.

Increased population, geographical mobility, and internal dissension made New England appear to be more like communities of convenience than organic entities; and by the mid-eighteenth century these trends were given further legal recognition when colonial charters contained not only stipulations for proportional representation but also requirements for periodic reapportionment based on censuses. This meant that territorial entities were officially recognized as convenient and contingent molds and not simply as organic entities. Proportional representation was to be an important element in the political philosophy of the Federalists and a part of the Constitution. But before examining the evidence from the Revolutionary period, we should note that there were still other uses of territoriality in the colonial era.

Other territorial effects

Grants of territory were the predominant basis of colonial government organization but early on there were settlements without such grants. Two famous ones are Plymouth and New Haven. It is interesting that the non-chartered colonies or companies had eventually to claim territory and did so by 'purchasing' it from the Indians (see preceding remarks by Roger Williams). While the purchases may not have been seen as such from the Indian's perspective, the settlers believed that the territory was needed and in a form that other Europeans would recognize. Colonies that were granted land had less compulsion to settle with the Indians living within the grant because, from the European perspective, Europeans already possessed it. Territoriality was the preferred mode of governance and this mode engendered more territory. Formerly non-territorial colonies either had to claim land of their own or were incorporated into adjacent colonies that had made such claims.

The colonial period exhibits many of the territorial effects we would expect from an emerging hierarchical organization. It was used by government to define realms of control and hierarchies of responsibility. But it also had the inadvertent effect of creating mismatches and spillovers. It appears unintentional because no clear evidence exists that territoriality was used at the time to disguise sources of power. Of the many mismatches and spillovers, an obscure example concerning New England parish structure is worth describing simply because its effects were so clearly inadvertent.

As we noted, Massachusetts was concerned that parish and civil authority should coincide and that Congregationalism be the official religion. The clergy of the colony originally wanted to be supported by voluntary contributions from their parishioners. But because voluntary contributions were not forthcoming in sufficient amounts, by the 1650s, the government had to promulgate taxes to support clergy. Rates to pay clergy were set on a proportional basis.[59] Even those who were not Congregationalists were to be assessed to support both the civil and ecclesiastical functions of the town and in 1692 the support of town clergy by town assessments was reiterated in the charter granted by William and Mary. By 1728, exemption to these taxes were granted to Anabaptists and Quakers, and in 1734 to Baptists. But there were no exemptions for those loyal Congregationalists who did not want to support their particular church or minister and yet who wanted to remain Congregationalists. They could of course found another church but they would still have to pay taxes to the church of the parish in which they resided. The only means they had of escaping a double tax was to convert to one of the religions exempted from supporting the parish church. The core of this paradox is that while territorial control partially succeeded in formally assigning domains of responsibility for support and for Congregational

Church attendance, it did not succeed in containing the people it intended to control.

The colonial territorial system had important consequences for government and for the conception of communities, but it should not be forgotten that the primary intention of colonization was to provide wealth to the mother country. Making grants territorial helped the mother country exploit the resources of the New World. To help foster exploitation, a significant portion of early charters contained long lists of resources that were claimed within the territory. These lists are surprising in light of the fact that territoriality could make it possible to state simply that a claim was being made to anything of value within the boundaries of the grant, and thus invoke the territorial advantage of classifying by area and not by kind. But perhaps there was need to remind the colonizers of the items in demand and perhaps it was the custom then to name what was of value rather than to use an abstract term such as resource.[60] In any event, the lists remind us that colonization was an economic venture at a time when Europe was entering a global commerical economy. Naming things of value to be taken from the land makes it clear that the land was valuable primarily insofar as it contained resources and commodities. But commodities change as markets change. Possessing land is speculative in a market economy. The changing nature of commodities and their fluctuations in value fosters the notion that space and its places at one time can be filled with things of value, and at another time, be void of value; that space is merely a framework in which commodities are located; and that place and events are contingently related.[61] This notion is reinforced by the unencumbered means of owning property, and by the abstract metrical system of describing property and political territory. Together these present space as an abstract system, conceptually and often actually emptiable and fillable.

Revolutionary period and westward expansion

The sixteenth to the eighteenth centuries – during which North American exploration and colonization took place – was a period of economic and political change for the rest of the Western World. Nation states were being forged and much of Western Europe was becoming the focus of an emerging global economy. It is not surprising that this period also saw the development of important new ideas in politics. Philosophers like Hobbes, Locke, Rousseau, Bodin, Montesquieu, and their contemporaries undertook systematic analyses of the nature of society and government. They were interested in what societies were actually like, and what they could or ought to be like. Collectively their inquiries raised several interrelated issues which were to have a bearing on the development of American political organization.

Among the most important questions they raised concerned the best type of government. The major choices at the time were monarchy, aristocracy, democracy (direct, i.e., without representatives, or indirect, i.e., with representatives) or some mixture of these. Closely related to the issue of ideal government was their conception of the type of society that was or ought to be governed. Was this community (to be) divided along social and economic classes and if so what kinds? Was this community (to be) large or small in size and population? What were its predominant social factions?

Each philosopher described somewhat differently the conditions of both the actual and ideal. Yet most thought that the type of government should match the type of society. There were grim theories such as Hobbes' which focussed primarily on explaining and justifying the concentration of power in a monarch.[62] The sovereign's power was transferred to him by the mutual consent of his subjects in order to prevent them from pursuing their unbridled self-interest and engaging in prolonged conflicts. Society without government was chaos. It was in a state of nature and would remain so unless people gave some of their authority to a sovereign.

In comparison with this 'state of nature' any royal abuse of power pales. Most others saw more complex relationships between society and government. They believed it was important to have the interests of the society represented (organically, that is, until the eighteenth century) and not to have too much power concentrated in one single office. Balancing power would temper it, make it just, and legitimate it. There was the specific but practical consideration of how the society should be subdivided and represented, of how territorial divisions and hierarchies were part of the systems of power. Some even considered how these parts and their functions were affected by the size and scale of the society.

It was generally held that the large states with numerous people as in England and France contained more factions, interest groups, or classes than did small ones such as city states. When the society was large and contained classes such as an aristocracy, a king, and 'the people,' their separate interests should somehow be reflected in government; perhaps by having the monarch voluntarily delegate to communities of interest degrees of independence (as with Bodin) or by having more formal or constitutional divisions of power.[63] This could mean dividing legislative power between aristocracy and the people while having executive power reside in the monarch (as with Montesquieu). Others who saw society as a multiplicity of interests believed some of these interests were localized in space, and that government ought to maintain a balance among local units. This could be accomplished through some form of geographical representation. Milton and Hume, for example, saw a division of representatives based on geographical scale. They urged local councils in each county to address questions of common interest and the national council to deal with the remainder.[64] There were some like Rousseau who believed the community ought to be small and thereby relatively homogeneous with regard to class interests.[65] This would make the will of the people become the interest of the group and would make possible pure democracy (government democracy without representatives) which was to be preferred to indirect democracy. States that were large in population and area had their advantages but they were not uniform in interests and could not have true democracies.

Founding Fathers

These and other strands of political theory were selected and transformed by the Founding Fathers to form new views of government to fit a new society. On the eve of the writing of the Constitution, there was some consensus about the issues. It was generally believed that a large political territory and large population offered some advantages, especially in the areas of foreign relations and defense, but was difficult to govern democratically. It was thought that over-concentrations of power must be avoided – but how? It was held that different interests or factions of the society should be represented – but what were they and how? Democratic government was the best form, but were its virtues attainable only in small communities? In addition, regardless of whether or not government was mixed or power was balanced, at that time only two kinds of territorial states were known. On the one hand were the unitary states wherein sub-territorial organizations like counties and towns derived their powers and very existence from the national or central authority, an existence which could be revoked if the central government chose. On the other were the confederations formed by the union of autonomous and independent territorial states which relinquished virtually no authority to the confederal government or league. The latter characterized the colonies and their union during the Revolution and under the Articles of Confederation. Before the writing of the Constitution no clear intermediate position between confederal and unitary or national form was known. The primary problem for the Founding Fathers, then, was to find a means of balancing interests and preventing concentrations of power for a large territory, democratically governed.[66]

One alternative, to continue to have a confederation with some modifications, was the position favored by the anti-Federalists in the state-ratifying conventions, many of whom saw no other solutions because they believed that large and populous nations could not be truly Republican. They argued that a large nation attempting to be democratic would eventually come to have a unitary form of government and that those who governed it would have too much power and thus be a threat to democracy. Representatives would be too remote, and representation too impersonal and aristocractic to be responsive to the people, all of which would increase the likelihood of strong factions taking over the reins of power. Separation of functions at the national level was not a sufficient check and balance. The only means of preserving liberty and yet attaining advantages of size would be to restrict the concentration of power at the top through some form of confederation – very much like the one under the Articles of Confederation.

The opposite view (expressed in the Virginia plan, for example) was to favor a unitary or national state democratically governed through indirect representation. It held that the only way of attaining the advantages in size was for the states to relinquish sovereignty to the central authority. To prevent over-centralization of power, the functions of government (legislative, executive, and judicial) would be separated into branches and would provide checks and balances over each other.[67]

The common ground of both positions was their shared desire for democratic government, their fear of over-concentration of power within the hierarchy, and their view that American society did not or ought not contain an aristocratic class. The opposition to the unitary plan came not only from the desire of anti-Federalists or Confederalists to have states retain sovereignty because they were already existing entities with interests, but from a genuine suspicion of indirect democracy in large-scale organizations and a concern that dangers from concentrating power would not be diminished only by dividing functions.

The profound and revolutionary compromise position came primarily from a few Federalists. They held that it was mistaken to believe true democracy flourished primarily in small territories. Small societies, contrary to conventional belief, they argued, were more likely to be faction-ridden and tyrannical than were larger ones. Madison in particular viewed American society as composed of factions; some, like manufacturing, agriculture, the rich, and the poor, were likely to last, while others were more ephemeral. Most important, factions or classes did not and would not necessarily correspond to territorial entities. Territorial partitions were in a sense like impersonal molds for a fluid geographical system. As Madison described it, 'these classes [within an area] understand much less of each other's interests and affairs than men of the same class inhabiting different districts.'[68]

The larger the scale of the society, the greater the chances that no faction could control the government while the converse was true, the smaller the scale.[69] It was not the large but the small states that threatened democracy. Having large areas and many citizens per representative, precisely because it increased impersonality, provided the representative with some freedom from provincial and factional interests. The concentration and power of factions could be reduced further by giving them various avenues or outlets to check and balance each other's influence. Territoriality was an essential means of providing such balances for them and for government in general.

Because territories did not coincide with the distribution of factions, territorial representation and territorial definition of constituencies, at different scales, would, if anything, fragment factions. For Madison, '*Divide et impera* the reprobated axiom of tyranny, is, under certain qualifications, the only policy by which a republic can be administered on just principles.'[70]

Though not a member of the Constitutional Convention, Jefferson voiced views about territory that were felt in the Convention and after. He believed that Americans shared common interests as an agrarian economy and that democracy could operate only through direct participation at the smallest possible level. He was especially concerned about the mismatch of territory and responsibility. To Jefferson, it would have been ideal if higher territorial levels were given only those functions that could not be addressed by the smallest units.[71] For the U.S. this meant that almost all matters were to be local or state except for defense and foreign affairs.

> It is not by the consolidation, or concentration, of powers, but by their distribution, that good government is effected. Were not this great country already divided into States, that division must be made, that each might do for itself what concerns itself directly, and what it can so much better do than a distant authority. Every State again is divided into counties, each to take care of what lies within its local bounds; each county again into townships or wards, to manage minuter details; and every ward into farms, to be governed each by its individual proprietor. Were we directed from Washington when to sow, and when to reap, we should soon want bread.[72]

Madison's views were the ones to prevail in the Constitution and his understanding of American society was to be the most prophetic. Yet Jefferson's detailed consideration of territorial hierarchy provided the mold used over and again to organize the new territories of the states. His proposals were incorporated in the Northwest Ordinance and Land Surveys, and his image of local authority and independence was to serve as a foil to the more realistic view of a strong centralized government and a society of factions.

Constitution

The prime distillate of the Founders' ideas about American government is the U.S. Constitution. This document uses territory to define constituencies for different parts of government. The unamended version stipulated (in Article I, Section 2): that members of the house were to be chosen

> every second Year by the People of the several States, and the Electors in each State shall have the qualifications requisite for Electors of the most numerous Branch of the State Legislature . . . Representatives . . . shall be apportioned among the several States which may be included within this Union, according to their respective Numbers . . . the actual enumeration shall be made within three Years after the first

Meeting of the Congress of the United States, and within every subsequent Term of ten Years.

The Senate was to be composed of 'two Senators for each State, chosen by the Legislature thereof' (Article I, Section 3). The President representing the entire country was to be selected by the process of electors. 'Each state shall appoint, in such Manner as the Legislature thereof may direct, a number of Electors, equal to the whole Number of Senators and Representatives to which the State may be entitled in the Congress.' The President is elected by the electors through a complex system of voting and tallying. (The process was amended on September 25, 1804, in Article XII.)

The understanding that these territorial units of representation were not entirely 'organic' units, but were at least in part convenient jurisdictions for mobile populations is demonstrated by the provision for proportional representation which would be adjusted every ten years according to the results of a national census. Proportional representation and the rules for reapportionment had begun to be incorporated in colonial government on the eve of the Revolution, and by 1780 all of what were once the original colonies elected their lower houses by direct popular votes and most had stringent districting requirements to equalize voting districts.[73]

The American system of government established between 1776 and 1789 may have been the first to conceive of its sub-units, the states, as generic territories – all alike in their form and place in government. No state name appears within the Constitution.[75] Although one state may have more representatives than another this can change again as population changes. No state is given a specific and unchanging role to play. Treating states as alike makes it simple for new states to be admitted. The sense that territories serve as convenient molds and are themselves subject to change is built into the Constitutional process of adding new states. Even the boundaries and numbers of existing states can change. Article IV, Section 3, states that:

New States may be admitted by the Congress into this union: but no new State shall be formed or erected within the Jurisdiction of any other State; nor any State be forced by the Junction of two or more States, or Parts of States, without the Consent of the Legislatures of the States concerned as well as of the Congress.

There are no Constitutional stipulations regarding the number of new states nor provisions for their size. The Constitution states simply that: 'The Congress shall have power to dispose of and make all needful Rules and Regulations respecting the Territory or other Property belonging to the United States.' (Article IV, Section 3) 'and guarantee(s) to every State in this Union a Republican Form of Government' (Article IV, Section 4).

The general procedures for carving new states out of the American West were outlined in the Northwest Ordinance of 1787 passed under the Articles of Confederation. This ordinance was based in part on Jefferson's 1784 plan for temporary government of the Western Lands ceded by Virginia to the Congress and for their division into states to enter the Union; in part on the ordinance of 1785; and in part on committee reports of May 10, 1786, September 19, 1786, and April 26, 1787. These ordinances and reports provided two territorial innovations. First, the territories in the Northwest were expected to be subdivided according to lines parallel to those of longitude and latitude, and these were to form components of the states' boundaries when feasible and the boundaries for practically all of the county, township, and private parcels of land. This rectangular land survey system was to be used subsequently through much of the West. Second, the Northwest territory eventually was to compose between two and five states. The exact number would in the end depend on the number of settlers in the region (Jefferson though had slightly different plans about the number and location of states, see Figure 1.2). This meant that not only were the populations in this territory to change but the exact number of states and their boundaries were negotiable. Forming new states was to involve a dynamic interrelationship between people and land. Once the state boundaries were to be created, a state would need 60,000 free inhabitants in order to be admitted into the Confederation. When the Constitution replaced the Articles of Confederation, the Northwest Ordinance was amended and adopted by Congress in the Ordinance of 1787.[75] Forming new states would generally follow the procedures outlined in the previous ordinances. Their boundaries would depend in part on the distribution of population. It was understood that boundaries of states already admitted into the Union still could be changed but only according to the procedures outlined in Article IV, Section 3, in the Constitution. All the states themselves had the authority to subdivide and alter territories within their boundaries, thus continuing at a smaller geographical scale the dynamics between people and territory.

The power to subdivide area into convenient units of 'counties, townes, burroughs, castles, manors etc.' as stipulated in the early grants and charters had now become the power of settlers to petition, and legislators to create, political units from states to counties, or townships, to cities or towns, and to dispose of sections and quarter sections of land; and most of these units were laid out in neatly nested geometrical territories. This was a system well suited for the times. Early nineteenth-century America contained a rapidly growing commercial economy with a developing industrial base. Population was burgeoning through natural increase and especially through immigration, and the country had vast lands at its disposal. Selling the land would immediately raise revenues for the federal government and in the longer run would increase the commercial and industrial base of the economy. A growing commercial and industrial economy requires freedom of geographical mobility for both labor and capital. Capital must be permitted to invest in different places offering different mixes of materials, labor, and government support while labor must be mobile enough to follow investments of capital. This means there must be a single set of laws regulating commercial activities over the entire country; uniform laws for banking, credit, currency, and so on. It means that property in land must be easily bought and sold. And it also means that political communities, from towns to counties, and states, must be ready to accommodate the mobile capital and labor to provide them with laws, schools, police and military protection, and with other basic services, and to do so without imposing restraints on their future mobility or on their connections with other places. These territorial units may differ in cultural contexts, in the degree to which they encourage capital or labor, but they cannot differ to the point where they obstruct the mobility of the two. The geographical conditions outlined in the laws and ordinances for subdividing territories, for admitting new states, and for surveying and subdividing land provided an indefinitely expandable and subdividable hierarchy of political–territorial units, each with some autonomy but all within a unified system.

Whereas these legislative enactments provided geographical containers for a transient population in a dynamic economy, they did not themselves define the balance between local autonomy and national integration, between the interests of the local farmer, the laborer, the merchant, and the industrialist. Politics at the national, the state, and the local levels were to cater to different interests in different regions, and these were to change over time. Moreover, due to mismatches and spillovers, the territorial forms themselves were to create problems of coordination; problems which plague any hierarchical territorial organization. The Constitution did fortunately provide guidelines regarding some aspects of state and federal relations. In addition to the enumeration of powers given to the federal government and the statement in Article X that 'the Powers not delegated to the United States by the Constitution, nor prohibited by it to the States, are reserved to the States respectively or to the people,' two parts of the Constitution have had an enormous impact in defining the interrelationships among territorial levels.

First, is Article I, sect. 8, beginning with the statement that 'The Congress shall . . . provide for the . . . general welfare of the United States . . .,' then enumerating the powers, and then concluding with the phrase in the last paragraph in Section 8, that Congress should '. . . make all laws which shall be necessary and proper to carrying into execution the foregoing powers.' The conservatives or strict constructionists interpret the last phrase narrowly whereas the liberals interpret it broadly. Lending weight to the latter interpretation is the Federalist number 31, explaining why the power of the federal government cannot be absolutely defined and limited.

Second, among the enumerated powers in Article I, is the interstate commerce clause (Section 8, clause 3) stating that 'Congress shall have the power . . . to regulate commerce with foreign nations and among the several states . . .,' which, in conjunction with other Constitutional authorities, has come to be used as a means of limiting the powers of state and local governments to enact laws that in any way restrict the movements of people and capital. Given the requirements of a dynamic, commercial economy it seems inevitable that these Constitutional provisions would eventually be interpreted and used to increase national integration by limiting local autonomy. Yet this was not clearly anticipated even by those among the Founding Fathers who saw the future of the United States to lie in trade and industry. Madison and Hamilton expressed concern that despite the balance of powers provided by the Constitution there was still a greater chance for the balance to swing in favor of the states. According to Hamilton, it

> will always be far more easy for the State Governments to encroach upon the national authorities than for the national government to encroach upon the State authorities . . . [because just as] a man is more attached to his family than to his neighborhood, to his neighborhood than to the community at large, the people of each State would be apt to feel a stronger bias towards their local governments than towards the government of the Union.[76]

The uses to which territoriality was put in the colonial period were repeated even more intensely at the time of the Revolution; especially those having to do with conceptually emptying and filling spaces, defining social relations territorially, and organizing complex hierarchies. The Founding Fathers assumed that their government would have a hierarchical territorial system. The problem was to avoid its pitfalls, and to use it to advantage. To some, the major concern was the tendency for territoriality at different levels of the hierarchy to create inequalities in knowledge and responsibility by giving those who had access to the top greater powers than those at the bottom. Once such inequalities exist in even the best-intentioned society they can corrupt those in power and limit the liberties of common citizens. Democratic elections may not be enough to constrain them because the size of the country and its numbers of citizens make representation remote and impersonal. Also, there was concern that responsibilities would be given to the wrong levels of government. Such mismatches of power and responsibility would create injustices and inefficiencies.

Not all saw territoriality's potential to circumscribe knowledge and responsibility primarily in a negative light. The Federalists believed the representative's remoteness from the local level to be of benefit. It gave the representative a loftier perspective and made him less susceptible to factional pressures. Federalists thought a principal means of ensuring liberty was to remove responsibilities from the local level, which was riddled with factions, and to give power to the federal government. Of course, even proponents of a strong national government agreed with the anti-Federalists in the need to be wary of too much concentration of power. It is in the way this concern was handled territorially that we find a great and original contribution – the use of territoriality's effect of dividing and conquering to check and balance concentrations of power.

The approach was original in its conception and in its detail. Whereas dividing and conquering or checking and balancing power has to be to the advantage of some group *vis-à-vis* another, those who were to be checked and balanced was never a matter of consensus among the Founding Fathers. Some saw the groups to be thwarted as factions, but others recognized that the states and federal government themselves (and those who controlled them) were the powers to be checked and balanced. Yet the role of territoriality was nevertheless understood and appreciated. Territoriality provided a versatile instrument for control, one which need not enumerate the individuals and groups to be circumscribed. Not needing to identify what is being controlled, yet empowering territories with varying degrees of authority and allowing for their multiplication, meant conceiving of territory in the abstract as a means of defining and molding social relations.

Again we should caution that our discussion has emphasized the use of the more modern effects of territoriality. These did not remove the 'older' ones, but were rather intermixed with them. We find in this period, as in previous ones, concerted efforts to keep the relationships between people and place intimate; to create the sense that community is more than a convenience. Late eighteenth- and early nineteenth-century American life was full of national patriotism and love of the land. This period cultivated a passion for America's boundaries and often portrayed America as a society rooted in the land. It was soon to be America's 'manifest destiny' to occupy the continent from sea to sea. Loyalty to place was directed also to the region, to the state, and to the local community. There were New Englanders and Southerners, Vermonters, and Virginians. Despite the abstract mechanical forms of settlement outlined in the Northwest Ordinance, and the rapid geographical turnover in this area, many of the first settlers still hoped to create closed organic units. Ohio, for instance, initially practiced the New England rule of calling out, but residency requirements soon had to be relaxed to accommodate the realities of geographic mobility.[77] Still the sense that people and place ought to be more than contingently related remained an important component in America's expectations about community. But the expectations were difficult to attain in a dynamic society. People became increasingly aware of the tension between the need to move and the desire to stay put. Rootedness became an ideal in opposition to rootlessness.[78]

Perspectives on twentieth-century territorial effects

The development of the nineteenth-century American political system is instructive in territorial uses. The South's secession provides a dramatic illustration of the complexities of territoriality. The Civil War deserves careful consideration from a territorial perspective but will not be explored here for, in the long run, it did not appreciably alter the direction of intergovernmental relations pointed to in the last section. The American political system in the nineteenth century continued to rely increasingly on two of the three 'modern' territorial effects stressed so far, namely: conceptually emptying space and using territoriality to create impersonal relations in complex bureaucratic structures. The general outline can be seen in the continuation of settlement in the West; the removal of the Indians to reservations; the increase in numbers of states, municipalities, and special purpose districts; the increase in governmental hierarchy and bureaucracy; and especially the centralization of power in the federal government and its executive branch.

These modern territorial tendencies, however, were to have somewhat different manifestations in the twentieth century. The Frontier was officially ended as a contiguous area in 1890; the literal empty spaces of America were filling up. Yet the expansion and intensification of capitalism made the modern territorial effects of conceptually empty space and impersonal bureaucracy even more of an integral part of the geographical environment. An increasingly interconnected financial system was penetrating every facet of life. To keep the economy growing, more goods had to be produced and newer types of products had to be invented. Producing commodities included not only physical things to consume but extended to services of all sorts and even to the commodification of leisure.

Anything and everything could be a commodity. Each product had a variety of forms and producers directed ever increasing attention to advertising and marketing in order to assure that consumers would desire a product and see it as different from another even almost identical one. As the numbers and types of products swelled, this increased attention to advertising and marketing gave business a degree of control over consumption and allowed it to be synchronized with production. The increasing penetration of the market both caused and was further assisted by technological and communication innovations. Geographical mobility of capital, of labor, and of communication spiralled. From the economic point of view, this had the effect of making places and space even more like commodities, and of molding society into masses who were simultaneously producers and consumers.

Although part of an abstract mass, people still differed in their characteristics and their social connections. From the economic perspective these differences were to be used to define markets for products, and new products were to create ever changing groups, sub-groups, and even individuals. Although we possessed the characteristics of age, sex, ethnicity, and although we possessed family ties and occupations, our place within these categories was more dynamic than ever, and through the commodities we would consume and the effects we would expect them to produce, we would be continuously presenting ourselves as new individuals, as belonging to new groups and associations. Both our geographical positions and our identities (or images) would become changeable. We would see ourselves more and more as potentially free and independent agents choosing our own locations, our own occupations, our own patterns of consumption. It is this heightened state of individualism and mobility in mass society that constitutes the significant context for examining the uses of territoriality in the twentieth century.

The first two modern effects – emptiable space and impersonal relations – are clearly evident in modern America. Even though the United States is literally 'filled' when compared to its use of land in previous centuries, the heightened geographical mobility and continuous commodification of place makes the public-economic meaning of space more and more a metrical system of locations and distances to which events are contingently connected. Within this context, political territories continue to be convenient molds for transient labor and capital. They are molds which can be conceptually emptied and filled, and these hierarchies of territorially defined communities reinforce impersonal relations. Moreover, the multiplication of territorial hierarchies and their ever more important role in organizing social relationships has increased the concentration of power in the upper levels of federal government. It has helped heighten the bureaucratization of government and has increased the power of the federal branch.

Theorists from the entire political spectrum would probably agree, if they were to have the issue presented to them, that the territorial effects of conceptually emptying space and of increasing impersonality and bureaucracy have intensified in this century.[79] Exactly what the indices are and what they mean would depend on which political philosophy one is using. Rather than consider how conservatives and radicals might trace the details of the historical relationships among territory, government, and society in the last eighty years of American history, we will take a briefer and more general approach and point out how the theoretical perspectives of each will lead to different evaluations of the overall processes. In terms of the theory of territoriality this means that the different political philosophies will especially emphasize different parts of the 'clusters' or effects, and interpret them differently.

Most political economists can agree that there has been a continued elaboration of the first two 'modern' effects of territoriality – conceptually emptying space, and advancing impersonality and bureaucracy. But most political economists would differ on the import they attribute to them and they would differ especially on the role they attribute to the third 'modern' territorial effect – using territory to obscure sources of power. In order to illustrate these differences in the evaluation of territoriality in twentieth-century capitalism we will examine how different elements of the political spectrum – namely neo-Smithians, neo-Keynesians, and neo-Marxians (named after the three influential political economists, Smith, Keynes, and Marx) – would be likely to address four interrelated issues in the development of American political territory if such issues were posed to them. The issues are: why some political processes are territorial; why they take place at different territorial scales; why there is a tendency for concentration at the largest scale, i.e., the national; and why there has been both a multiplication in the numbers of state and local levels and yet an increasing uniformity among them.

The names neo-Smithian, neo-Marxian, and neo-Keynesian may be more precise and less value laden than are the labels conservative, liberal, and radical. Yet they are still imprecise. When we refer to a neo-Smithian, for instance, we are discussing an individual who could probably not adhere to all of Smith's positions. The labor theory of value, for instance, is not likely to be held by any contemporary Smithian. An individual of course can hold positions that are Smithian, Keynesian, and Marxian.

These issues will be explored within the American context, but they can be recognized by Smithians, Keynesians, and Marxians as applying in part to other capitalist systems as well. This means that our approach to these issues and our review of the positions of the political right, center, and left are in part theoretical, addressing the overall relationships that can be expected between government and economy in capitalist societies, and in part specific to the American context, with its customs, laws, and Constitutional provisions for delegating responsibilities to different levels and branches of government. These four related issues will be presented first from the neo-Smithian and neo-Keynesian sides and then from the neo-Marxian perspective.

Why territory?

The neo-Smithians see government in capitalism as playing the role of umpire in the capitalist market system. Government clears the field for the game of capitalism and makes sure the players abide by the rules. Neo-Smithians view the market mechanism much as did Adam Smith. Each individual will pursue his own self-interest and through the invisible hand of the market will come the advancement of all. Yet neo-Smithians realize that the market system cannot provide everything. It needs to have the field cleared and the rules of the game enforced.[80] In modern economic terms this means that there are certain goods or services called 'public goods' which only government can provide. Neo-Keynesians agree that government must provide public goods, but the extreme neo-Smithians' position confines government exclusively to this realm.

'Pure public goods' are approximated in services such as law and national defense. For a public good to be pure it needs to possess three characteristics.[81]

1. First is joint supply. This means that the supply of a particular quantity to any one person does not diminish the possibility of supplying the same quantity of the good at the same price to any and all others.
2. Second is the impossibility of exclusion (non-excludability). This characteristic has two parts.
   a. It means that the supply to any one person prevents the good being withheld from any other person wishing access to it.
   b. It means that people who do not even pay for the good or service cannot be excluded from its benefits (or ills). This is often referred to as the free rider problem. The other side of the free rider problem is the tendency for people to 'underreveal' their preferences for this kind of good.
3. Third is the impossibility of rejection (non-rejectability): this means that once a service is supplied it must be fully and equally consumed by all, even those who might not wish to do so.

A public good is *impure* to the extent that one or all of these are not met.

Characteristics 2 and 3 comprise a public good's *externalities*. This means that those people who do not want the good, or who underreveal their preferences because they expect it anyway, will still be affected by its provision. Externalities can be positive or negative. A trivial case of a negative externality (but not of a pure public good) might arise if I buy a 'loud' shirt and wear it to the office. I purchased the shirt because I liked it, and the store sold it to me because they made a profit. The store and I – the two parties to the exchange – were satisfied. But when I wear the shirt, people might be offended. If they are, they would be affected by the product even though they were not parties to its provision, and each person's discomfort does not diminish the discomfort of others.

To the neo-Smithian, government's role would be to limit massive negative externalities and to provide massive positive ones. The chief reason why this falls mainly to government is that providing 'pure' public goods for large numbers of people requires enormous amounts of money. Private companies either would not have sufficient capital or could not expect to receive an adequate return on investment if they alone were to provide the good. Specifically, because of non-excludability, it is rational (i.e., in each person's interests) not voluntarily to contribute to the provision of this good (i.e., purchase it) because in the long run it would be provided anyway. Non-participation for an individual would be rational even when the good is 'impure,' if enough of the benefits would be coming to him once the good becomes available and if his lack of participation would not appreciably hinder the likelihood of its provision. This rationale of not participating is often referred to as a problem in the logic of collective action.[82] This means that it may appear to be in the self-interest of an individual not to participate in large groups that are after even an impure public good because the consequences of his individual effort would be small and his non-participation would save him time and/or money and would hardly be missed, while the participation of the other individuals would ensure the provision of the good. This is a problem of course because everyone else would think the same thing and no one would voluntarily participate.

For the most part, to ensure the provision of public goods means that there must be some kind of 'coersion' or 'additional incentive' to cooperate. In some cases this could be accomplished by distributing some private good to only those who contribute. More often it falls to the state to 'coerce' people to contribute or to ensure their participation.

Neo-Smithians and neo-Keynesians can agree that some public goods approach (if not actually attain) purity and that states need to help provide them. (They disagree on how often this occurs and on the degree of help.) This in itself would explain the role of government in capitalism if not its increasing size and functions. But our first concern is to focus on why government's role is *territorial*.

Still concentrating on the failure of the market places as the justification for government, we find that government's role in providing public goods territorially is explained in two ways. First, it is pointed out that externalities tend to be contiguous in geographic space.[83] But this contiguity in space is more likely a result of the fact that public goods are provided territorially. The second reason is therefore more fundamental. It argues that public goods are provided by political territorial units because they can levy taxes to support them while attempting to contain or exclude externalities and free riders. The reasons for the territoriality of public goods is, then, the reason for the territoriality of political units. It offers a means of defining and molding a community.

Levels of territories

For the nearly pure public goods like justice and defense, the territory would be the nation state. But in many cases public goods could be delivered more effectively if the nation were divided into local administrative units. It simply may be more efficient in a large country to have many federal courts dispersed geographically, each with its own area of jurisdiction, than to have all of the trials held at one place. The logic behind establishing these and other types of administrative or special purpose districts leads to the possibility of tiers of government, simply to delegate responsibility. Purely on the basis of efficiency, it could be argued that different public goods would tend to be provided most efficiently at different geographical scales. (In technical terms they have different economies of scale and different ranges and thresholds.) Taking only this into account, and disregarding the fact that the United States, for example, has tiers of government already in place and protected by the Federal and State Constitutions, it could be argued that an efficient form of provision would be to have separate or single purpose jurisdictions for each public good. Thus there could be a police district, a fire district, a sewer district, a school district, and so on. This logic of proliferation though is soon exhausted.

Many services are not completely independent of one another and can benefit by sharing boundaries with services that might theoretically have somewhat different geographical ranges. Fire protection complements police protection, and both may help school districts. It may pay, therefore, to have them all share boundaries in a multi-purpose district. Moreover, the existence of many special purpose districts would make it more difficult to hold public servants accountable. Hence the advantage to proliferating single purpose districts must be weighed against the advantages of decreasing their number through consolidation. Simply then for the sake of efficiency, a unitary state will have to have several territorial tiers to dispense national public goods. How many is open to question. In the case of the United States, the Constitution has guaranteed that smaller levels in the form of state governments will exist and that these have the power to tax, to elect officials, and to dispense their own local public goods. States in turn can subdivide their units and delegate powers to them as they wish. As a federation, the United States contains a political territorial hierarchy by law as well as by economic convenience. But the union of the two is not always smooth.

Providing public goods, often at several geographical scales, is the heart of the domain conceded to government by conservative political economy. By itself, and strictly interpreted, it does not provide much for government to do except supply an environment conducive to capitalism; yet the limited role it envisions can expand as the national economy expands. It should also be noted that whereas this limited role is based on the theoretical view of neo-Smithian political economy, actual business interests have not resisted government provisions of private goods when it suited them. Neo-Keynesians, of course, agree that the government needs to provide public goods, but they would also believe that the obligations of government ought to encompass many more impure public goods and that capitalism itself must be assisted by the intervention of federally funded expenditures.

Increasing role of national territory

Having public goods provided by a hierarchy of places, each trying to exclude 'free riders' and contain 'externalities,' often leads to a host of problems caused primarily by the fact that a territory does not always contain what it was expected to contain. In economics, the uncontainable consequences of an action are called 'spillovers' and these create a further need for larger scale governmental coordination. For example, all may agree that the provision of an adequate education to our youth is in the national interest. Suppose 'x' is the amount such education would cost per pupil. According to neo-Smithian theory, education would be private except for the truly needy who would have to rely for its provision on some kind of local governmental level. If responsibility for the education of the poor were left entirely in the hands of local territorial units (school districts, or even counties and states) there would be a tendency to under-provide for public education for the poor. Funds would be derived from local taxes, and, on the one hand, poor areas would find it difficult if not impossible to secure sufficient taxes to reach level 'x.' On the other, the rich would not be interested in paying for the education of others when they were also paying for the education of their own. Furthermore, if they were to pay for such education generously or even adequately, they might attract even more poor into their territory and increase their own tax burdens. The only way the poor could be adequately provided for would be by having (at least part of) the burden assumed by a yet larger political territorial unit, and this may lead to a greater role of government than neo-Smithians would wish.

Having a larger political territorial organization embrace and coordinate smaller ones may be the only means of assuring cooperation and efficiency among the local units. The absence of such cooperation could lead to sub-optimal provision of public goods, and these would soon appear as negative externalities at the national level.[84] The following shows how difficult cooperation can be without the help of a higher authority.

> . . . two jurisdictions may both invest jointly in a needed public facility with the result that the facility will be provided, and any economies of scale attained in this investment will make it cheaper for each jurisdiction. However, if one area invests independently it can attain the facility, but at a higher cost. The dilemma enters in that, without the collusion or cooperation between jurisdictions, the individually rational action for each government is to invest independently. The same logic applies also to tax competition. Two jurisdictions may both have strong incentives to raise local taxes, but this action undertaken by either jurisdiction independently will favor the other jurisdiction by encouraging . . . [out-migration]. Again, without

collusion, either the services of each jurisdiction will tend to deteriorate, or both jurisdictions will fall increasingly into debt . . . This [problem applies especially frequently] to the urban context in which large numbers of jurisdictions abut each other.[85]

Again, whether or not this kind of problem will be incorporated by the neo-Smithian model is an open issue. Yet a logical means of addressing it is to involve higher level territorial-governmental units. Such issues, as well as others created and aggravated by territorial partitioning, would be justifiable functions of government in the neo-Keynesian political philosophy. The neo-Keynesian view would see government not only as a provider of public goods and as a regulator and facilitator of the private market, but also as a social engineer and as an arbiter of intergroup conflicts.[86] Neo-Keynesians would see many more of the problems of collective action, externalities, and spillovers as constituting issues of national concern. Indeed, problems of unequal geographical distribution of resources are often created or aggravated by territoriality and to neo-Keynesians these issues would become appropriate concerns of the state.

Neo-Keynesians even more than neo-Smithians would expect increases in the size of government and both would attribute it to the general increase in complexity and scale of society. Moreover, both would expect the top of the government hierarchy to assume a greater role. In the U.S. case this would mean a shift in power from the states to the federal government, and, within each level, from the legislative to the executive, because the higher territorial levels and branches have access to more and broader information and have broader constituencies. (The neo-Keynesians may desire this shift while the neo-Smithians may see it as the only means to achieve just access and distribution.) Because states have more limited responsibilities, their policies are short-ranged when compared to national ones. States and localities cannot formulate their own plans without knowing what the national policy will be, while the national policy rarely takes into account the plans of the local community.

Increased numbers and uniformity of territories

Even the restricted role of government envisioned by neo-Smithian theory would result in territorial partitioning and hierarchy and lead to the Janus-faced relationship of territorial multiplication and integration or uniformity. The Constitution provides the legal foundation for both territorial proliferation and uniformity in the United States. With regard to the multiplication of the territories, the Constitution (Article IV, Section 3) expects new states to be formed and admitted. These were to be fairly regular with respect to government because they had to be modeled on the constitutions of one of the thirteen original states. The levels and authorities of substate jurisdiction are to be left to each state to decide. But unlike the relationship between states and nation outlined in the Federal Constitution, any arrangement by a state to subdivide itself into smaller jurisdictions would be a delegation of state authority which at any time could be altered or revoked by a state. Despite such discretionary state powers there is remarkably little variety in the hierarchical levels and functions of the subdivisions among the states. Most states are subdivided essentially into two basic governmental levels; county, and city or town, although there are many of each and also special purpose districts which do not fit (easily) into either of the two.[87]

More territorial explanations of why there are smaller units also leads to explanations of proliferation. Ever increasing complexity and specialization results in the tendency for greater territorial partitioning in order to help provide goods and services. The neo-Keynesian view is skeptical of how just and democratic are the territories and their services. The neo-Keynesians point to the imbalance and inequities in wealth and access that characterize the territories. For example, they point out that the rich, not the poor, are the ones that would be able to afford to move to more desirable places. So, in this sense, only the rich have the 'vote.' Capital is more mobile than labor, and a community that has made public investments in an infrastructure can be left high and dry when capital finds greener pastures. The rich are more mobile, and their absence makes the community even poorer. We see this tendency at work in 'white flight,' in the general plight of the inner city, and in the removal of capital from the Snowbelt to the Sunbelt or to foreign countries. The neo-Keynesian reaction would be a sympathy for the dislocated and a belief that the federal government has the responsibility at least to ease the transitions.

The opposite side of the multiplication of territories is their increasing uniformity. These territorial units offering citizens and businesses 'advantages' by creating different geographic mixes of opportunities still are part of a single political economic system and have been subjected to ever more uniform standards and procedures to allow for the mobility and integration of activities. In this regard the range of local authority has decreased, and uniformity of standards has increased. Once again, the general idea that economic growth involves territorial hierarchical differentiation and integration can be used to 'explain' the tendency to integrate. Neo-Smithian theory has not had much more to say about this except to make appeals for as little government as possible and at the lowest levels, whereas big business and its political allies have often endorsed government's role in increasing uniformity. Nor has neo-Keynesian theory added to the explanation of uniformity beyond seeing national political uniformity as an antidote to local injustices.

Neo-Marxist perspectives

Neo-Marxian theory sees a different relationship between the political and the economic in capitalism and thus also a different role for territory. Marxist theory does not view the political as a neutral referee allowing the rules of free enterprise to operate. The political is not an autonomous unit, but nor is it linked in a very simple way to the economic.[88] The state is there to further the interests of capital but it also intervenes against capital and on the side of labor. Capitalism makes government on the one hand its handmaiden and on the other the champion of the system's victims.

The relationships are complex because of the existence of different types of capital and labor, and because different levels of the state have different links to different segments of each. Every time the state acts, either to help a portion of labor or a portion of capital, it alters and even aggravates the tensions between labor and capital. The seed of the conflict is in the exploitation of labor by capital. The neo-Marxist agrees that the capitalist state needs to provide public goods, but the economy, in its capacity as provider of private goods, creates ever increasing tensions and dislocations. As capitalism develops, these potential antagonisms grow and so does government's role in addressing, repressing, or displacing them. For instance, capital must be free to pursue lower costs. This may mean a freedom to move to different locations offering lower raw material or labor costs. But in moving, capital can leave a community high and dry. Also capital must constantly invent new products and make old ones obsolete. Buildings and transportation systems and the entire built environment are commodities, often bought as public goods with public funds. Capitalism constantly subjects these to obsolescence. The invention of a new form of transport may destroy a city, making obsolete its older forms of capital and its labor force. Contradictions such as these involve the state in a complex web of relationships between segments of labor and capital.[89] Yet it is always in the interests of capital, if not always in the interests of the state, to divert attention away from the fundamental problem: the conflict between labor and capital. It is no surprise that from the Marxian perspective, one major set of territorial effects to expect in capitalism are those related to obscuring sources of power and to directing attention away from their purposes; and so in the following discussion of Marxian views of political territory we will explore how Marxists would address the four issues pertaining to territoriality in twentieth-century North America, but emphasize their concern that political territory can be obfuscative.

Marxian theory would expect that the levels of state government have the effect not only of supplying public goods at an 'economical' level, which in the case of the United States means supplying at least those services outlined in the Constitution, but also of providing circumscribed geographical environments which reproduce the culture and mentalities of different segments of the labor force.[90] Living in an underprivileged neighborhood means that the public services one receives are lower than the average and that one's expectations about services might thereby be lower. It means that one might not aspire to a higher level of education, or to a better job. And this means that a pool of people will be 'produced' expecting lower paying menial jobs. The opposite is the case in higher rent districts and more affluent communities. Here people grow up receiving more and expecting more.

Local governments and municipalities can also give people the sense that they can participate and voice their interests. Yet this may be a false sense in that the power of local government to affect fundamental change is limited. This sense of participation can serve to legitimate government while meeting few of its citizen's needs. It can also serve different needs of capital. Businesses operate at different levels. While American Telephone and Telegraph may lobby the federal government, the construction industries seek help through their local state houses.[91]

Increasing the number of local governmental territories can divide and fragment the consciousness of the working class.[92] People may form allegiances to their neighborhoods and districts, rather than to their class. Moreover, the localities will compete among themselves for scarce resources, rather than confront the rich. Thus territory can direct attention away from the conflict between labor and capital and toward a conflict pitting one ward against another, the city against its suburbs, the Snowbelt against the Sunbelt, or workers in America against workers in Mexico. But merging smaller political units to form larger ones, as is the case in the metropolitan form of government, can also be to the disadvantage of the poor because it can increase the costs of political campaigning and political advertising. This would make it difficult for the poor to run for public office.

Territorial hierarchy allows for the mismatch of power and responsibility which can further obscure the use of power. It is possible to assign local units of government the responsibility to solve problems with which they cannot cope. The task of preventing crime might be assigned to the local jurisdiction yet the incidence of crime is directly related to the state of the economy which is a national issue. Misassigning responsibilities to territories can be a strategy for unloading unpopular political decisions. A conservative president may not want welfare to be a national responsibility. Giving local communities the responsibility of poor relief may give the appearance that something is being done to assist the poor when in fact the problem is national in scope and in cause.

As for uniformity, Marxists would argue that it is more easily attained in a unitary state in which the territorial divisions are primarily due to efficiency in administration and provision of public goods and services. But in the case of a federal system, and the U.S. in particular, lower level territories, i.e., states, were originally autonomous and Constitutionally retain a significant range of authorities. Hence, in this case, one would expect uniformity to be more difficult to attain. But the same economic forces that make such units and their subdivisions necessary and important, also require that they not be too dissimilar and in fact that they become more alike in order to facilitate movement of people, goods, and capital.

As we noted before, for the United States, the legal basis for uniformity, like that for multiplicity, can be found in the Constitution. In addition to Constitutional implications of uniformity among state constitutions (Article IV, Section 4) and the Constitutional assertion that national laws are the supreme laws of the land (Article VI), the Constitution has provided several devices for affecting uniformity. There are the enumerated and exclusive powers of the Congress (Article I, Section 8) including the authority to coin money, to regulate interstate commerce, and to collect duties. There are explicit restrictions on state powers and prohibitions against Congress treating states unequally. For example, there is the right of preemption, there is the assertion that 'no tax or duty shall be laid on articles exported from any State' (Article I, Section 9) and 'no preference shall be given by any regulation of commerce or revenue to the ports of one State over those of another' (Article I, Section 9). There are the specific restrictions on the states in Article I, Section 10. There are also the provisions that state jurisdictions shall not interfere with individual civil liberties. For example: 'The citizens of each State shall be entitled to all privileges and immunities of citizens in the several States' (Article IV, Section 2) and the provision in the XIV Amendment that 'no State shall make or enforce any law which shall abridge the privileges or immunities of citizens of the United States.' According to Gordon Clark the interpretation by the courts of these and other provisions have stripped local government 'of any power that could meaningfully maintain the uniqueness and possibility of local autonomy [in the economic realm] . . . Political and economic power has become spatially and administratively centralized.'[93]

Marxists would emphasize that these and other legal bases have been used to the advantage of capitalism. Uniformity allows the individual – as consumer and producer – to be free of local territorial constraints. Uniformity permits the mobility of capital, labor, and products.[94] The general import is that the originally autonomous states in many respects have become little more than molds or receptacles for fluid capital and labor. This passive role of the local territories is ideologically justified as enhancing our freedoms of choice and movement. It did not take long for the forces of capital to use the Constitutional provisions to limit geographic differences and increase national uniformity. The degree to which the Constitutional grants have been interpreted in favor of centralization and uniformity and the degree to which the penetration of capitalism has extended the meaning of economic interconnection are illustrated in the legal interpretation of the 'Commerce Clause' and the right of Congress to mint money.

Originally the 'Commerce Clause' was thought to apply only to merchandise; however, the clause rapidly came to apply to the movement of all commodities, labor, and even information between and within states.

> Business transactions, whether formal through the exchange of money and/or commodities, or as implicit as in contract obligations that affected interstate commerce laws, have also come to be included within the scope of the clause's grant of authority . . . It also gave Congress the right to regulate interstate commerce . . . Congress can set the standards, conditions, prices and even the rates of commerce.[95]

The clause giving Congress the authority to mint money 'has been interpreted as applying to virtually all aspects of currency and exchange [including] . . . the velocity of circulation of currency, the form that currency takes and the uniform payment of debts.'[96]

These Marxist views concerning the role of sub-national territorial units can be extended to include the role of nation states within a global economy. In theory, nation states, unlike sub-states, are politically autonomous and, despite the efforts of the United Nations and regional alliances, there is, as yet, no global federalism, or confederalism approaching even the loosely centralized units of the original Thirteen Colonies. Nevertheless, international economic links have forged a degree of interdependence and cooperation which has tended to level national impediments to a global economy. National currencies are pegged, accounting systems are standardized, debts are generally honored, and, although tariffs are common, excessive interference in international trade will draw reprisals and international blockades are tantamount to declarations of war. Thus despite the absence of a central political authority, an international economic network has emerged that abides by rules that help to increase the flow of goods and capital and that diminish international economic differences.

Still, this system thrives on a degree of geographical difference provided it does not interfere with the mobility of capital and provided it offers to capital advantageous mixes of geographical opportunities. Thus one country may provide lower taxes, another lower wages, another political stability, and so on. By offering different social-economic packages, nation states then compete to attract a portion of the global economy. What is more, if the functions of the nation states are to create different geographical bundles that are conducive to capital, then these functions cannot be fulfilled unless nationalism inculcates loyalty to, and provides legitimacy for, these territorial units. In so doing, nationalism focusses attention away from the global level to the national one. It diverts attention from global relations among capital and labor to international and sub-national political issues. In other words, nationalism uses territoriality to obscure class conflict on the grandest geographical scale.

Applying the theory of territoriality to the political organization of North America has helped explain how and why political territorial units developed and the effects they have had on the American political process. All of the tendencies of the theory can be argued to be in evidence in the contemporary American political system but again the modern combinations are especially important. Political economic theorists of all persuasions would hold that capitalism's development of mass geographical mobility and commodification of place has intensified the sense that territory is conceptually emptiable and fillable, and that space is an abstract metrical system contingently related to events. The development of hierarchies of territories has defined communities to contain, channel, and mold the geographically dynamic processes; has heightened the effect of impersonality; and, in many cases, has increased the bureaucratization and centralization of power. Disagreement could arise concerning the evaluation of these effects and whether or not territoriality is used to obfuscate relations of power. The debate over these issues has taken the form of questions concerning the reasons for the territoriality of public goods, the numbers of territories and their geograpical scales, and the reasons for concentrating and centralizing authority. The entire political spectrum assumes that territory and public goods are functionally interconnected because territory is being used as a device to control and mold externalities, free riders, and to circumscribe the domain of users and contributors. Territorial hierarchy is also thought to be unavoidable in complex capitalist political systems because public goods have different geographic ranges and because there is need to coordinate spillovers and mismatches. Neo-Smithians add that the proliferation of territories resulting from the different ranges of public goods is also an expression of the freedom of a population to choose to control their own social environments.

Neo-Keynesian theory points out the dangers that come from territorial partitioning. It emphasizes the mismatches among jurisdictions, and how this exacerbates inequalities (a tendency frequently associated with territorial partitioning) and how territory can become an end rather than a means. But neo-Keynesians do not see these issues as having been created and used by capital to thwart the working class.

The neo-Marxist position extends the criticism and reformulates it by showing the negative effects of territory to be consistent with capital's interests. Territory becomes an (often unconscious) tool of the rich to exploit the masses. Territory helps to 'divide and conquer' the poor by fragmenting them into constituencies. It displaces attention from social conflict to territorial conflict and it mismatches to help avoid confronting socially sensitive issues.[97]

The political level is one of two realms intimately linked to modern uses of territoriality. The other is the work place. And, as we shall see, the development and interpretations of territoriality at this small or micro scale raises similar issues to the ones we confronted at the level of political territories. Indeed both levels of territoriality are intertwined.

6

The work place

Virtually every facet of life now occurs in a separate and distinct place set aside specifically for that activity. Factories have work stations for laborers, storage areas for tools and supplies, offices for secretaries and management. Schools have classrooms for different grades and offices for administrators, and the home has separate partitions for living, dining, recreation, and sleeping. The ill are kept in hospitals with wings and wards for different diseases and different needs. Criminals are in prisons containing cells and cell blocks to separate categories of offenders. The insane are in asylums. The elderly are in the old age homes. We go to museums to see great works of art; to symphony halls for concerts; to the theatre for plays. Streets are set aside only for movement, stores are places to purchase products, parks are places for recreation, and stadiums house organized sports. The list can be continued by adding types of factories, offices, schools, houses, prisons, asylums, old age homes, museums, symphony halls, theatres, streets, stores, parks and stadiums. But these are enough to suggest how most of our activities – indeed most of our lives – are so very spatially segmented and compartmentalized. Why has this spatial segmentation come about and what are its consequences? The answer is most clearly seen when we group the historical processes of the Modern period by their territorial effects.

Emptiable and impersonal space

These trends in segmentation manifest intense and minute territorial control. For work to occur as it does in a modern factory, for learning to take place as it does in a modern school, means that legal and social power must be brought to bear to exclude all other activities from a place, and to organize every last detail within it to serve the purposes of these activities. It is in specified places for activities such as work, recreation, learning, and healing that we experience territoriality most intensely. Yet we are so accustomed to these pervasive forms of spatial segmentation that we can forget not only that these are territorial units but that they are new ones at that.

Before the rise of capitalism, work, education, healing, and family were for the most part spatially intertwined. Laboring usually occurred at or near the home which for the vast majority was also the place of instruction. The sick and the elderly were not ordinarily physically separated from the rest of the community. Criminals were punished publicly by humiliation, torture, mutilation, or death; but they were normally not confined to prisons. Village and town streets contained multiple activities, they were not simply arteries for transportation. Merchandise was peddled in homes, in shops, and in public places, and entertainment occurred practically anywhere.

Things were not geographically neat and tidy in the pre-Modern World. Not everything had its place. Some of the flavor of this geographical admixture is rendered in contemporary paintings. A scene by Hogarth (Figure 6.1), for example, shows how untidy pre-industrial towns were. Lyn Lofland captures some of this intermixture in words when she presents her 'Composite Portrait of the Preindustrial City.'

Picture yourself dropped by means of a time machine, in the midst of some mythical composite preindustrial city. You are struck, first of all, by the sheer amount of activity there. Merchants, operating out of little cubbyholes, have spread their wares in the street. Next to them, a school of some type appears to be in session and you wonder how the students can concentrate. A wandering vendor is coming toward you, shouting out the wonders of his wares, and out of the hubbub of the street noises, you can just make out the shouts of other wandering vendors moving through the other streets. A man stops the vendor and they begin haggling over the price of an item. Then you notice that the more immobile merchants are also engaged in such haggling, some of it of long duration. Here and there, such an encounter has drawn a crowd and bystanders have involved themselves in the interaction. Everyone is shouting, with many insults passing back and forth.

Down the street, a beggar, his eyes obviously sightless, his face scarred by burns, keeps up a steady stream of requests for aid. As you pass through the streets you note that they are teeming with beggars. Some appear to be members of religious orders, others are lame or maimed, but many seem to have no bodily afflictions. Of the beggars, many are children; some are with adults, but large numbers 'work' alone.

You turn a corner and are struck with a sight repulsive to your modern eyes; a man has been nailed by his ear to a door. Upon inquiry, you discover that this is his punishment for cheating his customers. He is a merchant and he has been nailed to the door of his own establishment. Now a huge cry attracts your attention and you turn another corner to discover three people, stripped naked, being driven through the streets with whips and brushes. Later, at the outskirts of the city, you see the remains of bodies hanging in chains in some sort of elevated cage. One or two of the chains support whole bodies; from the rest, hang only 'quarters.'

Back in the city itself, a wandering actor or street singer or story teller has attracted a crowd and is in the midst of a performance. In one plaza, you note a public reader

Figure 6.1 Hogarth, *Night*
Reproduced by courtesy of the Trustees of the British Museum.

who is droning on and on about absolutely nothing and you marvel at the patience of his listeners.

Suddenly a town crier appears, shouting to all he passes that a fire is underway at such and such a place. You follow the crowd and, arriving at the scene, watch the citizenry attempt to cope with the flames. No one special seems to be in charge, and efforts to either contain or put out the fire appear hopeless. People are bringing things from their houses to help – bowls or buckets – running to the well with them and running back to toss the water on the flames. All is chaos. Eventually, after destroying many of the buildings in the area where it started, the fire burns itself out.

Watching the people running for water, you follow them and stop awhile at a nearby well or fountain, where you see men and women bathing, washing clothes, exchanging gossip. You see men with large buckets come by to fill their containters and then hurry off to deliver the water to nearby houses. And you see women coming with jars to get their own water.

The initial shock of this strange place having begun to wear off, you begin to note things that at first did not attract your attention. The sheer filth of the streets appalls you. Everywhere there seems to be refuse, garbage, human and animal excrement. And then you note for the first time the large numbers of animals in the streets – pigs eating the garbage, insects attracted to the excrement, dogs tearing at bones. There are larger animals too, horses and oxen drawing carts through the crowded, narrow streets, and other animals being driven to the marketplace. And everywhere children: playing in the streets, running up and down the narrow alleyways; taunting the lame and maimed beggars, poking sticks at an obviously feeble-minded man who is helpless to protect himself. Here and there they are joined in the 'games' by an adult, and your modern 'sensitivities' are stung by the discovery that human beings can find the suffering of others so amusing.[1]

This social and geographical untidiness was characteristic of most pre-modern cities, but, in Northern Europe, and then in North America, a major geographic change occurred with the rise of capitalism. The spatial changes from pre-modern to modern at this local or micro scale parallels those at the political or macro. At the micro level we again find the emergence of the three territorial effects associated with modernity: emptiable space, impersonal/bureaucratic relationships, and the possibility of obfuscation. Indeed the geographical processes of both levels are intertwined. Yet the two have exhibited important differences in territorial development. One concerns the empty space effect. At the macro level, an abstract metrical space soon became a part of the public's geographical consciousness. The discoveries of the New World allowed this system of space to become applicable to real areas, and, through political power, the empty spaces and territories of the New World could be filled with people and communities.

At the local or micro level the territorial effect of an empty space developed differently. Instead of starting with an abstract spatial grid which could fit a vast and conceptually empty space like the New World and then placing things in it, the micro level, beginning with the often crowded intermixture of events within a place, had to be 'thinned out,' so to speak, so that more and more places became containers for just one type of thing. When enough things were thinned out (despite the fact that the thinning out required the delineation of more and more territory), an abstract metrical spatial structure – and emptiable, fillable, and partitionable space (illustrated in Figures 6.2 and 6.3) – emerged as the underlying geograpical form. Thus the space of politics started as 'empty' and was 'filled in' whereas the space of work and leisure was 'thinned out' and then 'emptied.'

In both levels we find a multiplication of territories and hierarchies and the territorial effect of fostering impersonal/bureaucratic relationships. The abstract, empty space itself is felt as cold and impersonal. Yet some of the effects are experienced with greater intensity at the micro level. Here territorial control can penetrate to the smallest detail. A typist in an office is not only anchored to the work station, but the territorial control over this small space specifies even the physical orientation and inclination of his body. Whereas micro space can appear cold and impersonal, it is at this level that we also find the most intense efforts to humanize space. The micro level includes the home which is usually the place of greatest ease. At this scale also are found other attempts to establish places of warmth and comfort even in the coldest and most impersonal environments. A worker at a work station on an assembly line may feel more comfortable in his usual station than in another, and prisoners try to 'humanize' even their prison cells.

Increasing territorial partitioning can mean increased specialization and division of activities. But it can also present problems of unification and integration. We can contain things without knowing what they are and we can fragment without recombining. These difficulties of territorial integration, mismatches, and spillovers can have extremely personal consequences. Working in one place, learning and playing in others, and coming home to a house in which distinct rooms are set apart for playing, socializing, eating, and sleeping may affect our ability to understand how our day is interconnected. The geographical segmentation of our own activities can segment our sense of self.[2]

As was the case at the macro level, the territorial transformations at the micro have been affected by developments in the political economy and interpreting these effects depends once again on ones view of capitalism. But instead of three distinct viewpoints only two pertain at this scale; the neo-Smithian and the neo-Marxian. The type of territorial unit which expresses the differences in interpretation most sharply is the work place.

Figure 6.2 Emptiable architectural space: columns and trusses – small scale

Source: Oswald W. Grube, *Industrial Buildings and Factories* (New York: Praeger Publishers, 1971). Reproduced by permission of Verlag Gerd Hatje, Stuttgart.

Figure 6.3 Emptiable architectural space: columns and trusses – large scale
Source: Oswald W. Grube, *Industrial Buildings and Factories* (New York: Praeger Publishers, 1971). Reproduced by permission of Verlag Gerd Hatje, Stuttgart.

Interpretations

The neo-Smithian would approach changes in the work place in the same way as he approached changes in political territorial organization.[3] He would argue that the rise of capitalism increased economic complexity and interdependence and particularly the degree of specialization and division of labor. As a process becomes more complex, breaking it down into ever simpler tasks increases its efficiency. Even simple tasks can be accomplished more effectively with specialization and the division of labor. Simple and repetitive tasks save time and energy and thus allow more to be produced. Further savings could be had if all of the processes could occur near one another or even under one roof. Specialization and division of labor also make for greater interdependence, and specifying and coordinating this interdependence is the job of management. These interconnections are most clearly outlined by Adam Smith when he describes the organization of work in the famous pin manufactory:

a workman not educated to this business . . . nor acquainted with the use of the machinery employed in it . . . could scarce, perhaps, with his utmost industry, make one pin in a day, and certainly could not make twenty. But in the way in which this business is now carried on, not only the whole work is a peculiar trade, but it is divided into a number of branches, of which the greater part are likewise peculiar trades. One man draws out the wire, another straightens it, a third cuts it, a fourth points it, a fifth grinds it at the top for receiving the head; to make the head requires two or three distinct operations; to put it on is a peculiar business, to whiten the pins is another; it is even a trade by itself to put them into the paper; and the important business of making a pin is, in this manner, divided into about eighteen distinct operations, which, in some manufactories, are all performed by distinct hands, though in others the same man will sometimes perform two or three of them. I have seen a small manufactory of this kind where ten men only were employed, and where one of them consequently performed two or three distinct operations. But though they were very poor, and therefore but indifferently accommodated with the necessary machinery, they could, when they exerted themselves, make . . . upwards of forty-eight thousand pins [in a day]. Each person, therefore, making a tenth part of forty-eight thousand pins, might be considered as making four thousand eight hundred pins in a day.[4]

Smith believes the lessons from the pin manufactory are generalizable. 'In every art and manufacture, the effects of the division of labour are similar to what they are in this very trifling one . . . The division of labour . . . sofar as it can be introduced, occasions, in every art, a proportionable increase of the productive powers of labour.'[5]

Smith argues that the advantages of the division of labor are due primarily to three circumstances. First, there is an increase in the dexterity of the workmen. Second, there is 'a saving of time which is commonly lost in passing from one species of work to another.' Third, the division of labor has occasioned the invention of a vast number of machines which facilitates labor enabling 'one man to do the work of many.'[6]

Accompanying increasing division of labor comes the increasing need to coordinate economic activities and to place interconnected ones as close together as possible. This need by itself can make factories, rather than dispersed cottages and shops, the more economical spatial form. In addition, Smith points out, division of labor helps to increase technological inventiveness. The rise of capitalism saw the burgeoning of new equipment. Large and expensive machinery came to replace simpler tools. Machinery had to be housed in factories. And as more and more factories emerged and competition increased, it became even more important for the capitalist to organize work within the factory. The ever increasing scale of operations and the complexity of tasks made territorial partitioning and hierarchy an essential component of work. Places needed to be cleared for work to take place and only one type of activity could occur at a place. This territorial thinning out of activities increased with the heightening requirements of technology, complexity, scale, and timing.

It is important not to overlook the fact that the process occurs within the context of economic classes. One group owns the sources of production and control and supervise work, and another are simply workers, selling their time and working on machinery that they do not own. But according to a Smithian the relationship in the long run is fair. Workers may leave to find other jobs that pay more and owners/managers may fire workers and hire cheaper ones. Furthermore, the categories 'worker,' 'manager,''owner' are not separated by immutable class boundaries. With skill and good fortune a worker may become a manager or an owner, and, with poor skill and bad luck, owners can lose their fortunes. The harder the workers work and the more effective management becomes in organizing work, the better off everyone will be.

In this light, the place of work – the factory – appears as a necessary but neutral space cleared so that essential activities can be molded and coordinated to make everyone prosper. Even the impersonal character of the large factories, their hierarchies of territorial units, their large spans of control, and clear demarcations of responsibilities, add to the efficiency of work. Personal attachments would interfere with objective business and management decisions. Territorial circumscription of tasks increases the social distance among workers and between them and management. It helps make tasks clearer, simpler, and makes easier the training and replacement of workers and managers.

The neo-Smithian picture is, then, of a natural progression of territorial thinning out, specialization, and impersonal relationships to accompany increased social and economic scale, and complexity and division of labor. Although the neo-Marxist would also recognize empty space and impersonality/bureaucracy as effects of modern territorial organization at the micro level, he would explain them differently. Because he sees class conflict as the basis of capitalism he would naturally expect to find this conflict at the very core of the economic process – the work place. To the neo-Marxist, the relationships between labor and capital are unequal and oppressive, though the inequality is often disguised by the labor contract. This gives the appearance that the worker is free to sell his labor to any capitalist, just as the capitalist is free to hire labor. But in reality, Marxists argue, the worker must sell his labor and is not able to choose in any meaningful way.

The factory itself is not then a neutral place in which complex work occurs, but is rather a territory under the control of capital, to be used by capital to extract surplus value from labor. This can be done by literally locking labor out if it will not work for cheaper wages and for longer hours, or, as in the nineteenth century, literally locking workers in the factory from dawn till dusk. Capital's control of the factory permits it to segment the work process and restrict the movement of labor within the factory. Workers who once had knowledge of the entire process will no longer need it to accomplish their highly specialized tasks and unskilled laborers can replace formerly skilled ones. Territorially partitioning tasks may even create the illusion that there is more to the industrial process than actually exists. This illusion will make workers feel inferior and lead them to believe that they cannot manage the work process by themselves. Moreover, territorial partitioning isolates the worker from other workers and prevents him from comprehending the true relationships between labor and capital. In other words, the territorial subdivision and hierarchies of the work place are used by capital and management to divide, conquer, and de-skill the work force. Finally, the neo-Marxist would add that in light of the actual effect that territorial partitioning has of isolating the worker, capital's assertion that territorial partitioning is a natural and mechanical consequence of complexity and technology points out the very important role that is being played by the third modern territorial effect – that of obfuscation.

Again we must be careful to make clear that these are hypothetical positions and that one need not be a Smithian or a Marxian to hold some or even all of them. Even staunch Marxists can accept some neo-Smithian positions, and vice versa. Engels, for one, points out that specialization and hierarchical integration were indeed 'natural' accompaniments of modern technological life. With regard especially to the need for hierarchy he said:

> If man, by dint of his knowledge and inventive genius, has subdued the forces of nature, the latter avenge themselves upon him by subjecting him, in as far as he employs them, to a veritable despotism, independent of all social organization. Wanting to abolish authority in large-scale industry is tantamount to wanting to abolish industry itself, to destroy the power loom in order to return to the spinning wheel.[7]

These differences in interpretation of the modern effects of territoriality are best illustrated through the development of the work place. But the relationships between capitalism and the work place also affected the nature of a host of other institutions such as the asylum, the hospital, the prison, the school, and the home. And in these we find operating the same territorial tendencies as well as the possibility of alternative interpretations of their meanings. The evolution of these forms and their interconnections is complex; within the work environment, the timing of particular changes vary according to a host of factors such as the type of manufacturing process, the type of labor pool, and the role of government. Similar complexities pertain in the other institutions as well. Yet it is possible to rough-out the rudimentary changes in a few of them and have these stand as points of comparison for other studies.

Effect of work

The vast social disruptions accompanying the dissolution of feudalism and the rise of capitalism presented society with a need for new types of social organizations and controls. The uprooted peasants, the armies of vagabonds, and the burgeoning urban populations strained conventional systems of work, welfare, and personal forms of authority. There were vast segments of the populations that were in 'transition' – that did not fit conventional social categories. A significant reaction to these strains was to undertake geographically to isolate and contain the social misfits or deviants. No one has contributed more to the understanding of this important process than Foucault. From his works we see that the modern hospitals, asylums, poor houses, and prisons had common roots in the attempts of sixteenth and seventeenth centuries physically to isolate the increased numbers and kinds of the socially uprooted. Foucault called the seventeenth century the beginning of the great confinement and, according to him, 'what made confinement necessary was an imperative of labor.'[8]

Categorizing deviance as finely as we do now, so that we distinguish among physical ailments and mental ailments, among those who cannot support themselves because of social handicaps, physical handicaps, mental handicaps, and among those who will support themselves only through crime, has taken several centuries to evolve. The sixteenth- and seventeenth-century view of deviance was coarser: those whom we would now separate into different categories and place in different institutions – the criminal, the poor, the mentally ill – were often placed together in the same crude institutions of confinement.[9] Confining such diverse peoples together in often overcrowded and sub-sanitary conditions immediately presented the need for a typology of deviance that would help to subdivide and physically separate the confined so that they could be more effectively controlled. The problem required that the spatial organization of confinement permit the deviants to be carefully observed and the results of different mixes noted. Confining and organizing deviants became the responsibility of specific government agencies and spawned new ones as well as groups of private professionals. There was, however, an overriding criterion used to develop the typology of deviance, and that was work.

It is not surprising that work played a central role because many of the problems were to do after all with the changing nature of work. The transformation from a peasant to a commercial economy meant that a traditional communal economic interdependence was being replaced by a dependence on an impersonal market economy (see Figure 3.7), one which attempted to increase profits, decrease costs, introduce new products, and substitute one worker for another. To a considerable extent deviance was an inability to cope with such changes. It therefore seemed natural that the typology of deviance should be guided by the extent to which people were unable or unwilling to work and the reasons for their disabilities should be explained in terms of the characteristics of work. Thus the able bodied poor could be separated from the poor who were too ill to work, or from the poor who were mentally defective and could not attend to the drudgery of work, or from those who would not do an honest day's work.

The possibilities of work as a criterion for a typology of deviance was elaborated with the rise of industrial capitalism. Here, as illustrated in the comparison of 'a' with 'b' in Figure 3.7, the capitalist is not simply trading goods produced by craftsmen and cottagers working on their own tools, in their own establishments, and on their own time. The capitalist now actually is engaged in organizing and managing every detail of the working process. He can do this because he owns the tools of production, and work occurs in his own territory – the factory – and under his supervision. Within these conditions, specialization of tasks multiply. Work becomes minutely subdivided and often monotonous. The rigors of the factory refined the typology of work so that more categories and degrees of deviance could be isolated.

Work was used not only to define deviance, but to correct it. Under capitalism and the Protestant ethic work was a virtue.[10] It instilled habits of responsibility, discipline, and thrift. Forcing people to work would cure them of their disabilities. Prisoners were put to work, and so too were many of the ill. They would be healing themselves through their labor, or in failing to work they would reveal further their deviance. Even when houses of confinement became specialized as prisons, as asylums, as orphanages, and as poor houses, work was to be part of the regimen. '[Confinement's] repressive function was combined with a new use. It was no longer merely a question of confining those out of work, but of giving work to those who had been confined.'[11] Even when it was clear that classes of deviants were simply unable to work, the institutions in which they were placed were often organized like factories. Inmates were separated into clearly demarcated spaces so that they could be closely supervised. Their days were regimented. Activities followed rigorous time tables. The staff was hierarchically organized and the institution was to be run with factory-like efficiency.[12]

From simple sheds and buildings to contain people, these institutions of confinement, just as factory floors, became architecturally sophisticated purpose-built structures to classify, contain, order, and integrate. Hospitals became transformed through spatial and temporal regimentation from warehouses of the sick to institutions of moral reform.[13] Schools, too, were shaped by their mission as vehicles to educate the masses to work in factories. William Temple, an eighteenth-century English merchant, when calling for the placement of poor children in work houses where they should be set to work and given two hours of schooling a day, was explicit about the socializing influence of the process: 'There is considerable use to their being, somehow or other, constantly employed at least twelve hours a day, whether they can earn their living or not; for by these means, we hope that the rising generation will be so habituated to constant employment that it would at length prove agreeable and entertaining to them.'[14]

Workers valued education for their children but saw the acquisition of knowledge rather than the inculcation of factory discipline to be a school's primary mission. The differences in expectations were clearly put in 1861 by the assistant commissioner for the colliery areas of Durham and Northumberland:

> Time for school attendance is spared only with a view to its being preparation for work. Parents have no idea that there is any advantage in children spending so many years in school if the same amount of learning can be acquired in a shorter time. In short, they regard schooling, not as a course of discipline, but only as a means of acquiring reading, writing, arithmetic, sewing and knitting as a preparation of the main business of life – earning a living.[15]

Territory and space

The transformation of work, the rise of the factory, and the development of prisons, asylums, hospitals, and schools were all interrelated. They each were modeled on rational and efficient forms of managements. They all required the minute and intense subdivision and integration of territory. The whole must be territorially partitioned. Each partition must contain one type of individual or process. Yet each must be integrated into the whole.[16] The underlying spatial 'metaphysics' behind the geographical thinning out of events to create an emptiable and impersonal spatial surface to contain, classify, and organize human actions is illustrated in Foucault's discussion of architecture and control.

The regimentation of individuals or 'discipline' in his words 'proceeds from the distribution of individuals in space. To achieve this end, it employs several techniques.'[17] First it requires the establishment of territory or 'enclosure' as he puts it – 'the specification of a place heterogeneous to all others and closed in upon itself. It is the protected place of disciplinary monotony.' As an example of enclosure in France, he describes the great 'confinement' of vagabonds and paupers; and the creation of '*colleges*, or secondary schools' after the monastic model. There were also the

Military barracks: the army, that vagabond mass, has to be held in place; looting and violence must be prevented . . . The [French] ordinance of 1719 envisaged the construction of several hundred barracks . . . [to ensure] strict confinements: 'The whole will be enclosed by an outer wall ten feet high, which will surround the said houses, at a distance of thirty feet from all the sides'; this will have the effect of maintaining the troops in 'order and discipline, so that an officer will be in a position to answer for them.'[18]

There were of course the great factories with their thick walls and iron gates; opened once each morning at a precise moment to admit the workers, and locked securely shut to confine them until a precise moment in the evening when they were disgorged.

Enclosure is sufficient if the object is simply to collect and isolate. But with industrialization and the factory system, the objects enclosed were to be further subdivided, molded, and ordered. The interior of the space must be worked minutely and flexibly. This is accomplished first of all through the

principle of elementary location or *partitioning*. Each individual has his own place; and each place its individual. Disciplinary space tends to be divided into as many sections as there are bodies or elements to be distributed. One must eliminate the effects of imprecise distributions, the uncontrolled disappearance of individuals, their diffuse circulation, their unusable and dangerous coagulation . . . It [is a] procedure, therefore, aimed at knowing, mastering and using. Discipline organizes an analytical space

into the basic elements of 'cells.'[19]

But the territorial partitions must be integrated into a functional system.

Functional sites would gradually, in the disciplinary institutions, code a space that architecture generally left at the disposal of several different uses. Particular places were defined to correspond not only to the need to supervise, to break dangerous communications, but also to create a useful space. The process appeared [most] clearly . . . [in hospitals, prisons, schools, and factories] . . . In the factories that appeared at the end of the eighteenth century, the principle of individualizing partitioning became more complicated. It was a question of distributing individuals in a space in which one might isolate them and map them; but also of articulating his distribution on a production machinery that had its own requirements. The distribution of bodies, the spatial arrangement of production machinery and the different forms of activity in the distribution of 'posts' had to be linked together.[20]

The territorial partitions and their contents are both definite yet flexible. The interior organization of space can change, territories can coalesce or be further subdivided. In the absence of machinery, the parts can be interconnected by a schedule. As one moves from one place to another one is changing rank or class according to a serialization of places. For example, the division of a school into classes means that as students change their grade in school they also change their location in space. Each student

moves constantly over a series of compartments – some of these are 'ideal' compartments, marking a hierarchy of knowledge or ability, others express the distribution of values or merits in material terms in the space of the college or classroom . . . The organization of 'serial space' [as Foucault calls it] . . . made possible the supervision of each individual and the simultaneous work of all . . . It made the educational space function like a learning machine, but also as a machine for supervising, hierarchizing, and rewarding . . . [The spatial distribution] might provide a whole series of distinctions at once: according to the pupil's progress, worth, character, application, cleanliness and parents' fortune.[21]

Enclosure, partitioning, functional sets, and the serialization of space; these are the underlying structures of an abstract impersonal architectural form to organize and control behavior. The clearest exemplification of these principles in one actual design is found in the well-known philosopher Jeremy Bentham's description of a Panopticon. Bentham's design is contained in a series of letters he wrote in 1787 while visiting his brother in Russia.

The letters plus a postscript constitute the Panopticon publication of 1791. Many of the ideas contained in the Panopticon had been in the air for some time, but never before integrated and propounded as a set of architectural principles. The comprehensive power Bentham was placing in the organization of architectural space is foreshadowed in the first portion of the lengthy title of the work: 'Panopticon; or, The Inspection-House; containing the idea of a new principle of construction applicable to any sort of establishment, in which persons of any description are to be kept under inspection; and in particular to Penitentiary-Houses, Prisons, Houses of Industry, Work-Houses, Poor-Houses, Manufactories, Mad-Houses, Lazaretts, Hospitals, and Schools: with a Plan of Management adapted to the principle.'[22] The comprehensiveness of claims for architectural power is expanded further on in the preface. 'Morals reformed – health preserved – industry invigorated – instruction diffused – public burthens lightened – Economy seated, as it were upon a rock – the gordian knot of the poor laws not cut, but untied – all by a simple idea in Architecture.'[23]

Bentham continues to extol the powers of architecture in Letter One:

It will be found applicable . . . without exception, to all establishments whatsoever, in which, . . . a number of persons are meant to be kept under inspection. No matter how different, or even opposite the purpose: whether it be that of *punishing the incorrigible, guarding the insane, reforming the vicious, confining the suspected, employing the idle, maintaining the helpless, curing the sick, instructing the willing* in any branch of industry, or *training the rising race* in the path of education: in a word, whether it be applied to the purposes of *perceptual prisons* in the room of death, or *prisons for confinement* before trial, or *penitentiary-houses*, or *houses of correction*, or *work-houses*, or *manufactories*, or *mad-houses*, or *hospitals*, or *schools*.[24]

The Panopticon is intended to maintain complete order and control. This is to be accomplished by designing the structure so that the inspectors, while removed from view, can both observe the institution's inmates at any time, and have them believe they are under surveillance at any or all times. By allowing the inspectors to see, without being seen, 'the persons to be inspected . . . [will] always feel themselves as if under inspection, at least as standing a great chance of being so.'[25] The key is the 'apparent omnipresence of the inspector . . . combined with the extreme facility of his real presence.'[26]

What does this comprehensive and powerful device for control look like? A sketch included in later editions and shown in Figure 6.4 is based on the following original description, in which Bentham selects the role of prison as the primary illustration of the Panopticon's functions.

The building is circular. The apartments of the prisoners occupy the circumference. You may call them, if you please, the *cells*. These *cells* [as many as 900 of them][27] are divided from one another and the prisoners by that means secluded from all communication with each other, by *partitions* in the form of *radii* issuing from the circumference towards the centre, and extending as many feet as shall be thought necessary to form the largest dimension of the cell. The apartment of the inspector occupies the centre; you may call it if you please the *inspector's lodge*. It will be convenient in most, if not in all cases, to have a vacant space or area all around, between such centre and such circumference. You may call it if you please the *intermediate* or *annular* area . . . Each cell has in the outward circumference, a window, large enough, not only to light the cell, but, through the cell, to afford light enough to the correspondent part of the lodge. The inner circumference of the cell is formed by an iron *grating*, so light as not to screen any part of the cell from the inspector's view. Of this grating, a part sufficiently large opens, in form of a *door*, to admit the prisoner at his first entrance; and to give admission at any time to the inspector or any of his attendants. To cut off from each prisoner the view of every other, the partitions are carried on a few feet beyond the grating into the intermediate area: such projecting parts I call the *protracted partitions*. It is conceived, that the light, coming in this manner through the cells, and so across the intermediate area, will be sufficient for the inspector's lodge. But, for this purpose, both the windows in the cells, and those corresponding to them in the lodge, should be as large as the strength of the building, and what shall be deemed a necessary attention to economy, will permit. To the windows of the lodge there are *blinds*, as high up as the eyes of the prisoners in their cells can, by any means they can employ, be made to reach. To prevent *thorough* light, whereby, notwithstanding the blinds, the prisoners would see from the cells whether or not any person was in the lodge, that apartment is divided into quarter, by *partitions* formed by two diameters to the circle, crossing each other at right angles. For these partitions the thinnest materials might serve; and they might be made removable at pleasure; their height, sufficient to prevent the

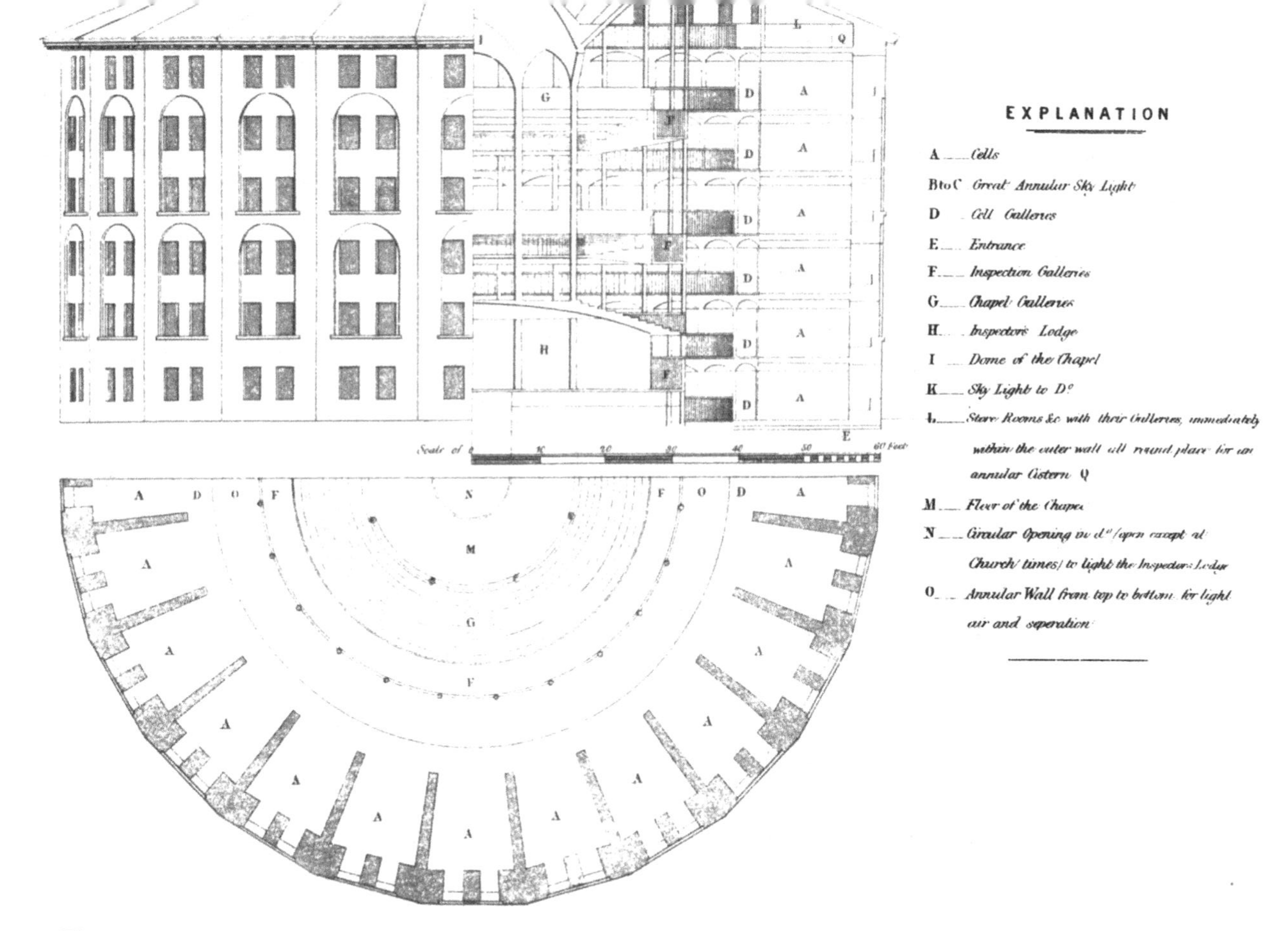

Figure 6.4 Plan of Bentham's Panopticon
Source: The Works of Jeremy Bentham, ed. John Bowring (Edinburgh: William Tait, 1843).

prisoners seeing over them from the cells. Doors to these partitions, if left open at any time, might produce the through light. To prevent this, divide each partition into two, at any part required, setting down the one-half at such distance from the other as shall be equal to the aperture of a door. These windows of the inspector's lodge open into the intermediate area, in the form of doors, in as many places as shall be deemed necessary to admit of his communicating readily with any of the cells. Small *lamps*, in the outside of each window of the lodge, backed by a reflector, to throw the light into the corresponding cells, would extend to the night of the security of the day. To save the troublesome exertion of voice that might otherwise be necessary, and to prevent one prisoner from knowing that the inspector was occupied by another prisoner at a distance, a small tin tube might reach from each cell to the inspector's lodge, passing across the area, and so in at the inspector's lodge . . . With regard to *instruction*, in cases where it cannot be duly given without the instructor's being close to the work, or without setting his hand to it by way of example before the learner's face, the instructor must indeed here as elsewhere, shift his attention as often as there is occasion to visit different workmen; unless he calls the workmen to him . . . or [makes use of these tubes]. They will save, on the one hand, the exertion of voice it would require, on the part of the instructor, to communicate instruction to the workmen without his central station in the lodge; and, on the other, the confusion which would ensue if different instructors of persons in the lodge were calling at the same time. And, in the case of hospitals, the quiet that may be insured by this little contrivance . . . affords an additional advantage.[28]

The key to management and organization is 'the centrality of the inspector's situation.' He possesses the most powerful device for supervision: 'seeing without being seen.'[29] Centrality, the transparency of the cell, and the invisibility of the inspector intensify the effects of supervision by giving to the inspector 'an apparent omnipresence.'[30] Territorial segmentation adds another economy to supervision. It permits one or a few inspectors to supervise many inmates and allows the under-supervisors to be themselves efficiently supervised.

A collateral advantage [the Panopticon] . . . possesses, and on the score of frugality a very material one, is that which respects the number of inspectors requisite . . . [and] another very important advantage . . . is that *under* keepers or inspectors . . . will be under the same irresistible control with respect to the *head* keeper or inspector, as the prisoners or other persons to be governed are with respect to them.[31]

Control can be absolute, pervasive, and minutely ordered, yet adaptable to a variety of functions. The building's interior partitions are flexible enough so that the number of cells, their sizes and their degree of partitioning, as well as the connections among them and the lodge, can be altered to fit different purposes. The type of activity to be contained determines the finer points concerning the internal spatial organization, and especially the intensity of control within (and the degree of permeability among) the cells. The cells 'will . . . be more or less spacious, according to the employments which it is designed should be carried on in them.'[32] When the establishment is to be used for a manufactory,

> [t]he centrality of the presiding person's situation will have its use at all events; for the purpose of direction and order at least, if for no other. The concealment of his person will be of use in so far as control may be judged useful. As to partitions, whether they would be more serviceable in the way of preventing distraction, or disserviceable by impeding communication, will depend upon the particular nature of the particular manufacture.[33]

But the minute organization of space, even for prisoners, need not necessarily be used to punish and coerce. Rather, by controlling what is present and absent in an environment, the Panopticon can be used to further health, virtue, and morality. The organization of space acts as a moral agent, containing the good and removing the bad. This is why Bentham believed the structure could be used not just for a prison, but for a school, separating scholars and grades of scholars, and also for a boarding school to 'fill the whole circle of time, including the hours of repose, and refreshments and recreation.'[34] 'All distraction of every kind, is effectively banished.'[35]

Bentham knew that it was more difficult to persuade the readers of the Panopticon's moral benefits than of its coercive effects. To allay the public's fears that the institution could be used only for coercion he takes considerable pains describing its suitability for such humanitarian purposes as hospitals. 'If any thing could still be wanting to show how far this plan is from any necessary connection with severe and coercive measures, there cannot be a stronger consideration than that of the advantages with which it applies to hospitals.'[36] In the remainder of the treatise Bentham expounds on designs of the building and its fixtures (including the types of furniture, plumbing, ventilation, and heating) that are most appropriate for specific functions.

Bentham's contribution was in uniting moral principles and architectural designs that were already coming into practice in the early and middle eighteenth century. His synthesis was both abstract and concrete. Although his designs were detailed, they were intended to provide a versatile structure, illustrating the potential for hierarchical territorial control. In the abstract, the Panopticon embodies principles of territorial organization that are eminently suited to the use of space by industrial capitalism. It presents a comprehensive architectural design for instituting the modern territorial effects of any emptiable, fillable space, and of facilitiating impersonal/bureaucratic relationships. The building is a shell containing an abstract volume of space which can be variously partitioned, emptied, filled, and interconnected. Within its walls, spatial subdivisions impose impersonal social relations, facilitate hierarchy and efficient supervision, and make space and events contingently related.

At the concrete level, the Panopticon design has inspired nineteenth- and twentieth-century institutional and factory architecture. Several attempts have been made to follow Bentham's plans literally and construct Panopticon prisons. Among the most famous were Robert Adam's design for Edinburgh Bridewell, Gandy's design for Female Felons at Lancaster Castle Gaol, Joshua Jebb's Pentonville, and John de Haviland's Philadelphia Penitentiary.[37] Many more nineteenth- and twentieth-century institutional structures have adopted Bentham's principles and some components of his design. Even though they may differ considerably in detail – especially in altering the overall shape from circular, to rectangular or square, and in combining shapes to form wings and units – and in functions – whether they be hospitals, asylums, prisons, or schools – their overall effect (as in Figures 6.5, 6.6, and 6.7) is of an architectural shell containing relatively flexible sub-territorial units to allow for surveillance, organization, and control.[38]

It was the transformation of the work environment which created the need for an architecture of surveillance and control in the first place, but it is the work environment for which the detailed – though not the abstract – principles of the Panopticon are least suited. Bentham did not pursue the fact that work processes and machinery and their spatial interrelationships are so varied that any complex geometric form, with even flexible inner partitions, may be too constraining to provide a general mold for work. Many industries such as steel smelting and hydroelectric generation need to be housed in unique spatial structures that are designed specifically for them. But there are other processes for which the machinery is not so monumental and there are many more for which the equipment is relatively small. Each of these processes may be containable within a general type of factory building. Thus an electronics firm, a sheet metal shop, a cardboard container manufacturer, and a printing company may simply require heated, lighted floor space, and therefore can each occupy the same type of structure. How they use the space within differs, but to all of them can pertain the abstract concept of a factory as a shell. It is for these types of work processes especially that the theory, though not the details, of the Panopticon apply.

Modern industrial architecture has in fact taken the principles of the Panopticon one step further. Partitions within the factory became far more flexible than Bentham anticipated, to the point where most were quite literally invisible in the form of work stations. Such a station may be defined as the worker's imposed personal area. It may have a center, perhaps a desk or piece of machinery, but it also has a boundary which, though not physically marked, is clearly etched in the mind of the worker and supervisor, and may well have been specified in the original drawings for the layout of the plant.

Anchoring workers to well-ordered work stations in a plant without physical partitions extends the possibilities of surveillance and control. A

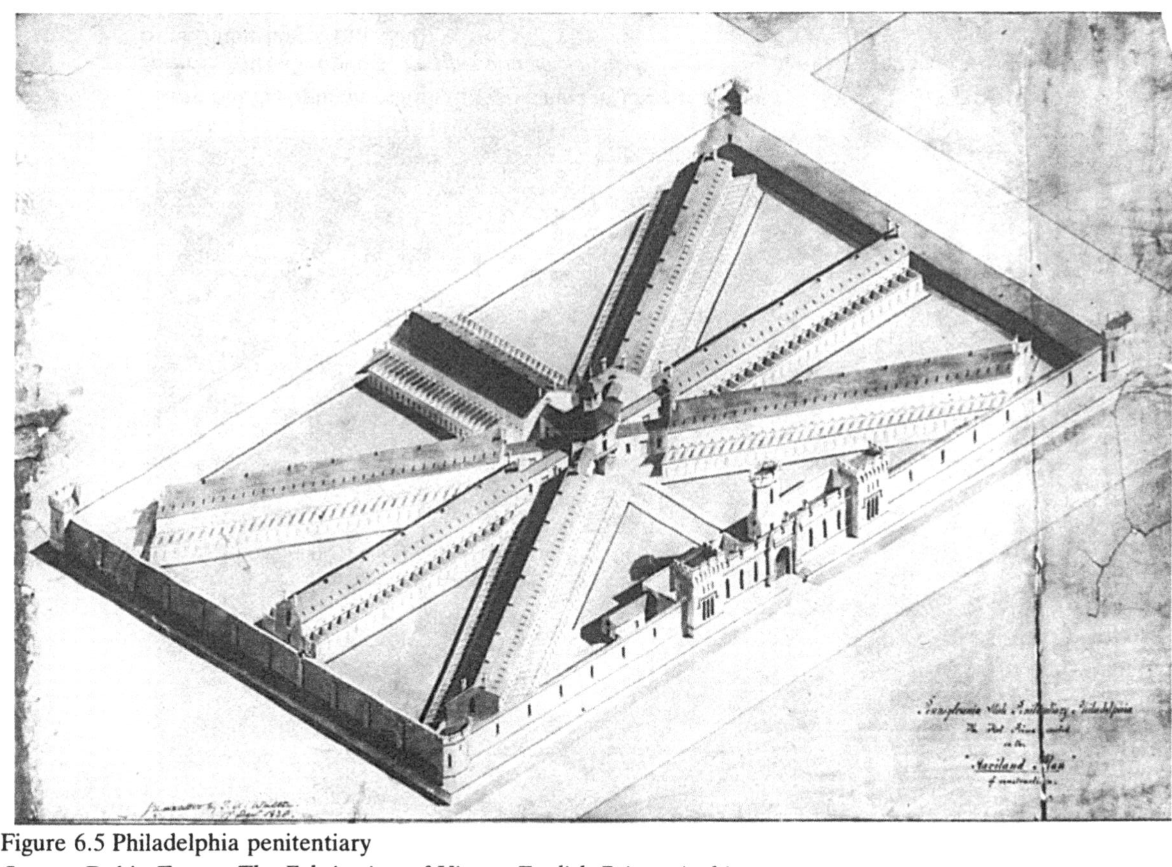

Figure 6.5 Philadelphia penitentiary

Source: Robin Evans, *The Fabrication of Virtue: English Prison Architecture* (Cambridge: Cambridge University Press, 1982).

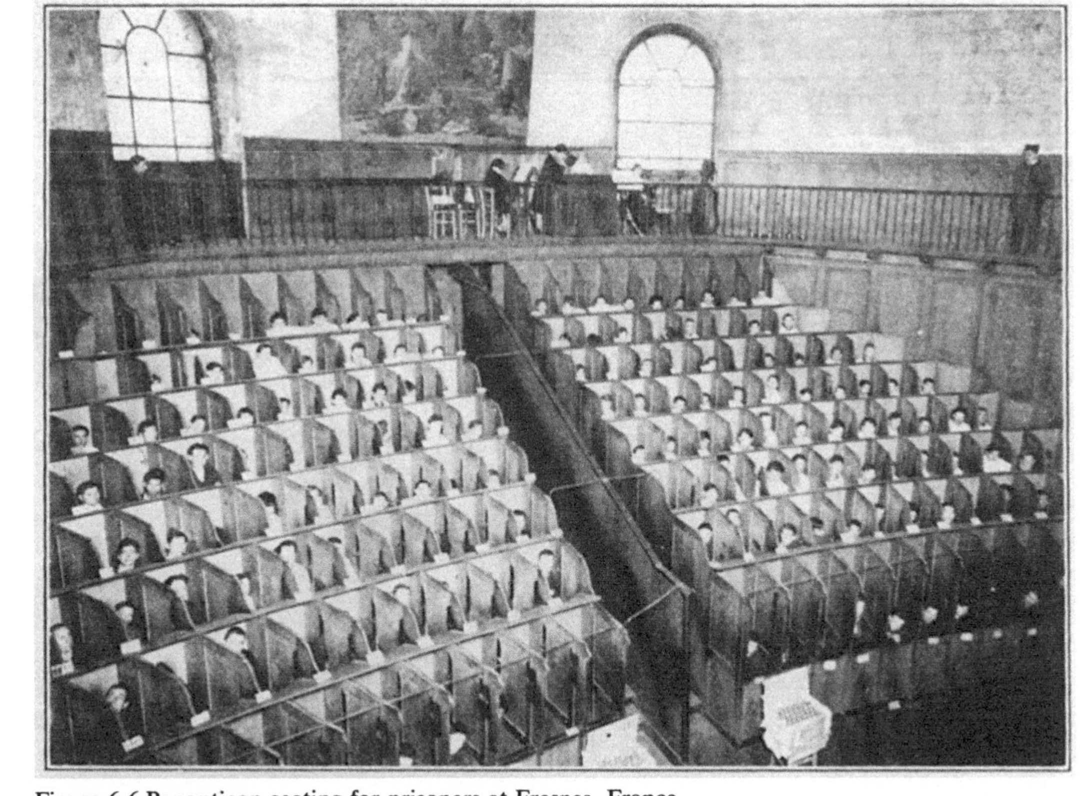

Figure 6.6 Panopticon seating for prisoners at Fresnes, France

Source: Alfred Hopkins, *Prisons and Prison Buildings* (New York: Architectural Book Publishing Co., 1930).

Figure 6.7 Interior view of Bentham's Panopticon as interpreted at Joliet, Ill.

Source: Alfred Hopkins, *Prisons and Prison Buildings* (New York: Architectural Book Publishing Co., 1930).

single supervisor, while walking from station to station, giving this worker or that personal instructions, can still be in the position to see, at a single glance, what is happening on the rest of the floor. And the worker too may see that the supervisor's gaze can encompass the entire floor at any moment. Foucault describes how these conditions pertained for the late nineteenth-century Oberkampf factory at Jouy.

> The largest of the buildings, built in 1791 by Toussaint Barré, was 110 metres long and had three stories. The ground floor was devoted mainly to block printing; it contained 132 tables arranged in two rows, the length of the workshop, which had eighty-eight windows; each printer worked at a table with his 'puller,' who prepared and spread the colours. There were 264 persons in all. At the end of each table was a sort of rack on which the material that had just been printed was left to dry . . . By walking up and down the central aisle of the workshop, it was possible to carry out a supervision that was both general and individual: to observe the workers' presence and application, and the quality of his work; to compare workers with one another, to classify them according to skill and speed; to follow the successive stages of the production process.[39]

Technological developments in building material helped to make interior space more pliable. The late eighteenth-century introduction of cast iron columns in functional architecture eliminated the thick and obstructive interior stone or wood pillars. Iron columns increased visibility, flexibility, and the sense of an uninterrupted empty space.[40] By the mid-nineteenth century, cast iron was used for the entire frame of many buildings, which further increased the size and flexibility of unobstructed interiors.

These advances in architectural engineering made it possible to utilize every bit of interior space for production, and with these advances space itself took on an ever greater economic significance allowing areas of space to be graded according to their degree of economic efficiency. Rent had been the customary monetary measure of the economy of space, but, in the nineteenth century, industrial space came to be described and measured also as productive or unproductive. Productive space was defined as 'the area of effort, the theatre of those operations which distinguish a factory from a vacancy.'[41] The opposite was called 'waste ground,'[42] 'waste places,' 'unremunerative space,' or 'unproductive territory,'[43] or mere 'vacancies.'[44] These 'have to be warmed and cared for, and walked around and through, without in any way adding to the productive powers of the shop; but, instead, actually decreases the output by the increased transportation always consequence upon unproductive territory.'[45]

At the end of the nineteenth century, steel construction in a particular format called 'skeleton construction' was commonly employed in office buildings and factories. Skeleton construction, pictures in Figures 6.8 and 6.9, raised to a fine art the possibility of having a large part of the floor space 'without partitions so that they could be subdivided later on to suit the

Figure 6.8 Skeleton construction

Source: Siegfried Giedion, *Space, Time, and Architecture* (Cambridge, Mass.: Harvard University Press, 1967).

Figure 6.9 Modules of space: model of the services parts processing redistribution center of the Ford Motor Company, Brownstown, MI

Source: Oswald W. Grube, *Industrial Buildings and Factories* (New York: Praeger Publishers, 1971). Reproduced by permission of Verlag Gerd Hatje, Stuttgart.

tenant.'[46] It is this interior flexibility that is the chief advantage of skeleton construction: 'the skeleton – whether iron or steel or reinforced concrete – is essentially a neutral spatial network. Its "cage construction" bounds a certain volume of space with complete impartiality, and no one intrinsic direction.'[47] This type of flexibility was a keystone in the architectural philosophy of one of the major industrial architects, Alfred Kahn. According to Oswald Grube, Kahn held that the 'layout of the factory should be sufficiently elastic to permit rearrangement in accordance with changes in production methods, or expansion of department or production expansion without disorganizing the exact scheme.'[48] In the United States especially, factory design generally includes the possibility for complete changes in the production process taking place within the building. Industrial architects themselves speak of constructing containers, volumes, spaces or 'connected sheds of gigantic size which can be altered fairly easily according to the changing demands for space of a company's various divisions or departments.'[49]

Home

We have discussed the territorial effects of an emptiable space and of facilitating impersonal relations in the context of institutions and work places. What of the home? Certainly the home is the most personal space of all. It is filled with content and meaning, and can hardly be compared to the interior of institutions or work places. Yet there have been architectural developments in home construction which suggest that the use of space has to some degree followed its use in work. The eighteenth- and nineteenth-century removal from the streets of all activities that interfered with movement and commerce, and the thinning out of functions within shops and factories, had their counterparts in home architecture. Before then, large houses and castles, the only structures with any internal partitions to speak of, may have had many rooms, but those for living were rarely places reserved for specialized functions. Food may have been prepared in any one of several compartments that contained a hearth, depending on the number of people to be fed; and the same room may have served also as a bed chamber and a social hall. Moreover, there were no passages or halls set aside for the sole purpose of moving from room to room. To get to one room usually meant going through another. It was not until the early seventeenth century[50] that the interiors of the grand homes became specialized and that passages were introduced.

As interiors of houses became more specialized, the partitions themselves became more flexible; more so in American houses than in others. American architecture has been marked by one tendency since the arrival of the earlier settlers. The American house has generally contained a ground plan which can be enlarged (and contracted) whenever new social and economic conditions make it desirable.[51] The New England continuum is the classical case of an expandable house place. Americans had no compunction about cutting through a house and dividing it into two or more dwellings. Houses were even freely moved from one lot to another. Assisting in this flexibility was the American 'balloon frame' construction, a wooden forebearer of the skeletal frame construction. It seems that Americans, more than others, used sliding doors as flexible partitions. Flexibility in interior design extended to apartments. In the late nineteenth century, at the same time office buildings were being designed with skeletal frames, apartments were constructed with removable partitions so that individual rooms could be joined together.[52] Indeed, flexible inner space is one of the great contributions of such modern architects as Frank Lloyd Wright.[53]

A house is not a factory, or asylum, or prison. Surveillance and management are not its explicit functions. Yet efficiency of movement is certainly a major consideration in house design. Kitchens are laid out as scientifically as are some factory work places, and corridors and rooms are spaced so as to maintain an efficient flow of traffic.[54] Efficient heating, ventilation, and lighting also play a role in house design. Allowing the interiors to be flexible makes the assignment of events and space more open. This means that there is an underlying consciousness on the part of the architect and the user that space and thing are only contingently related, that space can be repeatedly subdivided and reintegrated, emptied and filled. One can argue, though, that this is not a sign that space is abstract and impersonal because the flexibility permits the occupant to arrange and design the interior of the house as he or she likes. This allows one the opportunity to give the place one's personal stamp; to have it express one's personal tastes. Against such an argument can be raised the question of the sources of our personal tastes. How often do they come from others: from television, from the neighbors, from fashion magazines, or from interior designers? Are we expressing *our* tastes or are we defining those of *others* by the objects we consume? And how comfortable can we be with our interiors if we change them so often?

Obfuscation

We have sketched how the thinning out of architectural space – the use of territory to create a sense of an emptiable and fillable space and of impersonal social relationships – developed and reinforced the nature of social and work relationships in capitalism. Beneath the surface of this discussion have lurked the normative issues concerning this transformation. Modern abstract emptiable space and impersonal relations may in themselves appear neither to be benign nor malevolent. But have they been used overall to one or the other effect? We cannot hope in any way to prove that they have or have not; we can only present arguments favoring one interpretation or another.

Two comprehensive views that can be used to construct normative judgments about the processes we have investigated are the neo-Smithian and the neo-Marxian. Each would find in the transformation some mixture of good and evil, but the neo-Smithian would emphasize the former and the neo-Marxian the latter. The neo-Smithian argument for the micro level would be similar to that for the macro. He would contend that the uses of territory within capitalism can be of benefit to society as a whole. While perhaps thinking that Bentham's Panopticon is extreme, the neo-Smithian would hold that specialization and the division of labor do require segmented places and a hierarchy of supervision. As the former increase, so too do the latter, and the entire process is accelerated by the development of new technology, machinery, and the scale of operation. Overall, the territorial effects help increase productivity and efficiency which, under capitalism, is for the benefit of all.

The neo-Marxian, of course, would point out that the territorial uses and effects cannot be properly understood unless they are placed within the context of capitalist class conflict. Although the wage-labor agreement may disguise the conflict, there is an inherent inequality between labor and capital which is translated into a conflict over the work environment. Capital will use the work space to subjugate the worker and extract from him surplus value. The territorial subdivisions and hierarchies within the work place will be used to divide, conquer, and deskill the workforce. Calling the process a natural and beneficial outcome of technology and complexity as do the neo-Smithians would strongly suggest to the neo-Marxists that capitalism is employing territoriality to achieve the third modern territorial effect: that of obscuring class conflict.

It is important to remember that the territorial effects linked theoretically with obfuscation are the same as those that can be linked with the ordinary operations of complex organizations, but with a malevolent twist. The territorial division of knowledge and responsibility can be obfuscative if the effect is not true efficiency but rather, for example, to keep the majority ignorant of the entire process. Assigning a small-scale territory the responsibility for a process that spills over its boundaries may be an inadvertent mismatch, or it may be a means of deceiving others by claiming that something is being done about the process. Cases for the normative import of territoriality and especially for the presence of obfuscation can be presented most clearly if we first examine relatively simple examples of micro territorial changes. An ideal opportunity is provided by those shifts from cottage industry to factories which occurred *before* there were major changes in technology. In these cases people shifted from a condition of working with specific tools in their own homes to working with practically the same tools in a shed or factory owned and supervised by others. This is to say that there are important cases in which the development of the factory system was not dependent on the invention of large-scale machinery. Focussing first on such examples in which little or no new machinery was introduced and keeping these examples in mind as we discuss more complex cases allows us to isolate territoriality effects and present arguments about their normative implications.

At the beginning of the commercial revolution many producers owned their own tools and produced their products in or near their own houses. Work often involved the entire family. Even apprentices lived in the workers' homes as members of the family. These households were part of what has been called a domestic or cottage economy (Figure 3.7a).[55] The workers were producing a specialized commodity and were dependent on the market for their livelihood. They and their families often toiled for long hours under poor conditions – the houses were often meagre establishments. Yet the worker (and the apprentice when he became his own master) was still in charge of his own production process. He could decide to some extent when and how he worked within his own establishment.[56] He could if he pleased stay up all night and then take the next day off. This freedom of scheduling work and sharing the tasks with the family members allowed workers some opportunities for engaging in other activities to supplement their market incomes. With the exception of a few weaving factories, which will be discussed shortly, most of the textile industry in England was based on this type of 'domestic system' until the nineteenth century. Before this period most English textile laborers worked on their own looms in their cottages, and many possessed, in addition, some land for gardening and even an animal or two.[57]

These 'hand loom weavers' might sell their cloth in their own homes (if they lived in towns), or they might bring their wares to a central market place or a guild hall for cloth weavers, as was the case in parts of Yorkshire,[58] or, as was common in Leicestershire, a merchant might provide intermediaries called 'baggers' to bring to the cottage weavers wool or cotton which had been carded and spun in other cottages, and then to collect from them the finished products, the yards of cloth.[59] Several technological improvements did occur in the weaving industry, but, for the most part until the invention of the various steam powered looms in the late eighteenth and nineteenth centuries, the cottage weavers were able to construct, purchase, or else rent these new tools or machines and thus still work at home. While the cottagers were often impoverished and worked long hours for little pay, they still maintained control over the production process. They did not, except as apprentices, work in a place away from home and under the control of someone else.

In the case of weaving it was not the invention of large and expensive power driven looms that occasioned the rise of the first factories (although this technology did eventually put most of the cottage hand loom weavers out of business), for, as we noted, a few experiments with weaving factories began well before the invention of steam powered looms, in some cases as far back as the sixteenth century. While these early cases were rare they did leave a mark as prototypes of factory organization and discipline and they were to increase in number and importance as we approach the industrial revolution. What happened in the early cases was simply that someone with sufficient capital to buy a number of looms would do so. Perhaps the market was expanding and this person saw that there were unemployed weavers without looms; or perhaps there was a depression in the textile market, and cottage weavers, no longer able to find a market, needed to sell their looms to raise money to support their families. In either case the capitalist who purchased the looms stored them in a large building or shed, and when the market improved he had those workers who did not now own their own looms come work in his building under his supervision.

Among the most famous of the early textile industrialists are William Stump who is credited with placing over 100 looms under one roof, and Jack of Newbury who had even a larger establishment.[60] A book of verse, *The Story of John Winchcome*, published in 1597, describes the manufactory of Jack of Newbury. This establishment was supposed to have contained 200 weavers together in a large room managing 200 looms, assisted by 100 female carders, 200 spinners, 50 clippers and 80 dressers.[61] Importantly, these early factory owners were renowned for their management and disciplinary skills. Increasing numbers of cases of such manufactories or work shops can be found in other industries as well as in textiles as we approach the industrial revolution.[62] They can be thought of as belonging to part of a territorial continuum: from cottage, to cottage work shop, to manufactory, to large-scale factory.[63]

The critical point of comparison between cottage and manufactory is that the same products can be turned out on the same tools in either place. What, then, are the advantages of one over the other, and who profits by them? Most neo-Smithians would begin by pointing out that the factory would be more efficient than would be the cottage. Placing the process under one roof under the control of one party, economizes on movements, on schedules, and on internal agreements and exchange.[64] In the case of weaving it would synchronize weavers, spinners, and carders. It would eliminate the wasted transportation of wool or cotton and cloth from cottage to cottage. Moreover, having many 'hands' gives everyone the opportunity to specialize, and this, as Adam Smith has argued, increases efficiency by increasing dexterity and inventiveness. It also allows the tasks to be simplified to the point where less skill is required for each so that fewer skilled workers at cheaper wages can be hired and thus the price of the product can be reduced. To a neo-Smithian, these effects are to everyone's advantage. They will provide an affordable product and put poor people to work. These innovations cannot take place, however, without the organizational skills of the factory owner or manager. He must be able to subdivide the work process, to coordinate its parts, and train and manage workers. By providing these essential skills he creates jobs, and thereby not only earns but justifies his income.

The success of early 'industrialists' in all areas was often openly attributed to their managerial discipline even more than to their use of new technological developments.[65] Contemporary manufacturers themselves explained that placing the process under one roof would create efficiencies of movement and reduce expenses by having fewer intermediaries, if the workers could be trained and disciplined.[66] The critical role of management itself became formalized and subdivided by the nineteenth century into specialized fields. Although the entire factory process may seem a natural progression, it was not put into place without enormous struggle.

It was difficult to discipline a workforce for the regimen of the factory. Laboring in a factory was different from domestic labor, and workers often resisted the change, though the resistance varied according to type of labor and industry.[67] Labor's resistance to factory discipline was an important reason why the factory system did not sweep the textile industry before the use of steam powered looms. Much of the distaste for factory work came from those cottagers who saw the factory system's lower wages and cheaper products as threats to their own livelihoods. Such concerns were voiced as early as the 1550s in complaints against weaving work shops and manufactories which 'cheapened' labor.[68]

The inconvenience and problems that workers attributed to the early and small ventures into factory work became magnified as the factories increased in size and number, and new dangers became evident. By the nineteenth century, the factory system in all manufacturing realms was seen also as a threat to the fabric of family life.

> If the factory system prevail . . . it would call all the poor labouring men away from their habitation and their homes into the factories, and then they will be obliged to work separate, and they will not have the help and the advantage which they have at home. Supposing I was a parent and had four or five or six children, and one of them was fourteen, another twelve, another ten: If I was working with my family at home I could give them employment, one to wind bobbins, another to work at the loom, and another at the jenny; but if I go to the factory, they will not allow me to take those boys, but I will leave them to the wide world to perish.[69]

Another and old source of resistance was the fear that factory time and discipline would diminish the worker's flexibility. He would lose command of his own working schedule. Even when the cottage work was more uncertain, when pay was lower, and the number of hours the cottager devoted to work was longer than would be the case in the factory, he nevertheless seemed to have placed a premium on the freedom to schedule his labor and leisure within his own home. This freedom was both difficult to measure and was coming to be eroded as early as the eighteenth century with the enactment of legislation to allow authorities to enter workers' cottages to see if they had kept any scraps of cloth or other materials that were technically the merchants'.[70] Such losses of autonomy and the increasing pace of domestic production to many independent weavers still did not tip the balance in favor of factory employment.

The extraordinary lengths to which some would go in order to retain the 'freedom' of cottage work is illustrated in the measures taken by silk ribbon weavers in nineteenth-century Coventry, England. Relative to other areas of weaving, steam driven looms came late to fine silk ribbon weaving. Not until the mid-nineteenth century were power looms producing a majority of ribbon. The industry had seen the introduction of new technology in the Dutch 'Engine' and the Jacquard loom, but both of these were hand looms, and moreover in the productions of fine ribbon even these were inadequate. In the early nineteenth century practically all of Coventry's ribbon was made by hand loom workers in cottages. This was still the case by the early 1830s although the uneven work hours of the domestic weavers and the disputes over embezzlement between them and the merchants made the latter think highly of the factory system in which 'men and women worked regular hours, at a steady pace throughout the week' so that by 1838 several manufactories employed hand loom weavers in work shops and factories.[71]

Modern power loom factories for ribbon were introduced into Coventry in the 1850s and these were seen by journeymen working at home as an enormous threat to their incomes and freedoms. To remain at home these journeymen needed to compete with the modern system. The remarkable point was the steps they took to resist becoming employees in modern power loom factories. Many of these Coventry weavers were living in three story 'row' cottages; that is, each three story house shared its exterior walls with another cottage so that an entire block contained an uninterrupted row of attached buildings. The third story was an attic 'topshop' in which weaving took place. The cottagers' alternative to entering a modern steam powered factory was to let the 'monster' of steam enter into their houses; to turn their own homes into steam factories. What they did was to 'place a steam engine at the end of a row of weavers' houses, [and] . . . conduct the power up to the topshop at the end of the row, and . . . transmit the power by shafting down the row from one topshop to another through the partition walls separating the houses.'[72] By the early 1850s several hundred cottages had been thus transformed, and new ones were being constructed on this basis. There was even a scheme for developing small urban 'utopias' of several blocks based on these 'factory cottages.' Of course they could not compete for long. The expenses of renting the steam engine (most if not all were rented, not owned) made it necessary for each cottager to work long hours. He had to pay his share of the rent even if he were too ill to work. By the end of the 1860s these pressures reduced the cottage work shops in Coventry to a thing of the past.

Many of the fears of factory work expressed by these and other domestic producers were to come to pass once factories with power machines replaced cottage industry. Conditions within these early factories in most industries were often cruel. Men, women, and children worked for long hours, under miserable conditions, and for very little pay.[73] The association of factories with prisons, asylums, and poor houses was not based simply on the fact that work was thought to be a moral agent and that the factories served as an organizational model. The connection was more direct in that inmates from these institutions often comprised part of the factory workforce.[74] They did so in two ways. First the institutions themselves supplied work for their inmates, thus combining prison and factory. 'Some workhouse-manufactories were started at the end of the seventeenth century, including Bristol, in 1697 and Exeter in 1698–1701, but the main impetus came from the Act of 1723 which rapidly led to the building of at least sixty such workhouses in the provinces and fifty in London.'[75] Second, and more widespread, was

> the massive employment of pauper apprentices in private industry . . . These pauper apprentices were not employed because they were [necessarily] cheap . . . Apprentices were taken on because otherwise the mills would have been without sufficient labour or at least without sufficient child labour . . . [and] the pauper children represented the only type of labour which in many areas could be driven into them.[76]

Whereas there is no doubt that the territorial control of the capitalist over the work process in the factory reduced the workers' freedom, it seems impossible to determine if, overall, the factory system actually created more hardship and impoverishment for the working poor than did cottage industry.[77] The neo-Smithian would argue that if the working poor were emiserated it was a temporary setback, and that the standard of living of factory workers in industrialized countries since the early and mid-eighteenth century had risen considerably. And, if the workers lost some freedoms by entering the factory, they were eventually to gain more in prosperity and a shorter working day. These gains, of course, would be due to the general increase in efficiency that comes from the factory system.

The neo-Marxist would point out that the efficiency argument for the factory system obscures some critical distinctions. First, if there is an efficiency in working close together under one roof, then this same efficiency can be accomplished by having it take place within a workers' cooperative instead of a capitalist factory. A cooperative would allow workers who work together in one place to be in control of their own tools and to have a say in the management of their work processes. If the workers in a cooperative could use these same tools, on their own, in close proximity to one another to produce the same commodities as were produced in capitalist-run factories, then what does the owner and manager of the factory contribute to productivity and efficiency?

The neo-Marxist answer would be that the capitalist does not in fact contribute any essential role to the process. The managerially imposed efficiency described by the Smithians is a narrow measure, and often includes having the workers work longer hours for less money. Making labor work longer for less pay is how the capitalist reduces his cost per product and extracts a profit. The system is based on demeaning work and de-skilling labor.

In summarizing the entire process, Gras states that the move to the factory was

> partly for purposes of discipline, so that the workers could be effectively controlled under the supervision of foremen. Under one roof, or within a narrow compass, they could be started to work at sunrise and kept going till sunset, barring periods for rest and refreshment. And under penalty of loss of all employment they could be kept going almost all throughout the year.[78]

Echoing Gras, Braverman states:

> Control without centralization of employment [within a factory] was, if not impossible, certainly very difficult, and so the precondition for management was the gathering of workers under a single roof. The first effect of such a move was to enforce upon the workers regular hours of work, in contrast to the self-imposed pace which included many interruptions, short days, and holidays, and in general prevented a prolongation of the working day for the purposes of producing a surplus under then-existing technical conditions.[79]

In those industries that did not have expensive machinery, a deskilling of the workforce and a cheapening of the tasks of labor may have been the sole means by which a capitalist made a profit. If the processes, which were familiar to the craftsmen, could be spatially partitioned within the factory, and if the workers were not permitted to view the entire factory operation, but rather were kept spatially segmented, the workers might then think there was more to the process than there actually was. Whereas each skilled craftsman, on his own, could have drawn up plans, ordered raw materials, operated and repaired equipment, and stored the finished product until the time to sell it; in the factory each of these processes would be undertaken by specialists who are spatially restricted to their respective work stations, be it in the drafting room, the store room, the production floor, the warehouse, or any other such compartment. According to Marglin, 'Separating the tasks assigned to each workman was the sole means by which the capitalists could, in the days preceding costly machinery, ensure that he could remain essential to the production process as integrator of these separate operations into a product for which a wide market existed.'[80] Even in the relatively simple cotton factory, a manufacturer noted that a competitor did not allow any of his employees, not even his manager, to mix cotton 'so that he can never take his business away.'[81] *The Spectator*, 1866, as much as admitted that the role of management is artificial, when it pointed out that a cooperative, 'although showing that workmen could manage shops, mills, and all forms of industry with success, and [that the cooperatives] immensely improved the conditions of the men, [were nonetheless defective because they] did not leave a clear place for the masters.'[82]

The efficiency of the factory, which the neo-Smithian attributes to the capitalist work system, is seen by the neo-Marxist to be a narrow measure of production which incorporates pervasive, though often disguised, forms of exploitation assisted by territorial relationships of authority within the factory – especially the modern territorial effects of emptiable space and impersonal relations. To the Marxists, the spatial segmentation of labor, the restriction of movements within the factory, the territorial separation of long- and short-range planning, all help to deskill labor, to fragment labor's knowledge and responsibility, and to divide labor itself. Yet these effects are obscured by the rationalization of capital which presents them primarily as neutral geographical components to an efficient and impersonal process.

The development of improved power driven machinery and instrumentation further transformed the effects of territory. Under these innovations the factory worker is virtually an appendage to a machine. He is anchored to it in space. Even the orientation of his body and the movement of his limbs are determined by the needs of the machine. It is the power driven machine and not the worker that sets the pace of work. And this mechanical pace becomes an aid and a substitute for management's personal supervision, for it further intensifies and disguises the functions of supervision. According to Braverman: 'Machinery offers to management the opportunity to do by wholly mechanical means that which it had previously attempted to do by [supervisory] . . . and disciplinary means.'[83] Citing Babbage, Braverman adds 'one great advantage which we may derive from machinery, . . . is from the check which it affords against the inattention, the idleness, or the dishonesty of human agents.'[84] No longer need the supervisor be physically present at every moment to make sure that the workers are working to full capacity. Now the pace and duration of work can be set by the speed of the machine. Perhaps only an occasional glance by the foreman would be sufficient to allow him to judge if the worker was attending to the job. And, in many cases, even this glance may not be necessary, for many types of machines have instantaneous feedback loops which can warn a supervisor if the machinery is malfunctioning or if the worker is not attending to his task. Such machinery may even contain built-in tests for quality, thereby incorporating yet another function of supervision.[85]

Having machines assume part of the role of supervision opens up the possibility of new geographical relationships between workers and supervisors. No longer need the workers be placed along rows in a wide open factory space. Nor need there be as many supervisors actually present on the factory floor. With the introduction of modern telecommunications processes it is even possible that for some businesses workers need not be working together under one roof. The tasks may be carried out in smaller and more dispersed locations, and some even at home. The communications systems can allow information to travel efficiently from one unit to another and automatic feedback devices can monitor the rate and quality of work.[86] Neo-Smithians may herald such changes as increasing the workers' flexibility and freedom. Through these innovations workers may be able to exercise more control over the timing of their work and the nature of their work environments. The neo-Marxian may counter that these changes can make the work process more impersonal and mysterious, can further obscure the managerial functions of supervision and control, and further fragment the workforce and reduce class consciousness.[87]

The two political economic viewpoints provide important differences in interpreting territoriality under capitalism. It should not be forgotten though that these differences are consonant with territoriality's theoretically possible effects, and that the theory of territoriality pointed out that different political economic theories would emphasize some territorial potentials over others. Territoriality's logic is not a captive of one particular political economic theory. This makes it possible for different views to emphasize some territorial effects, and for other effects simply to be beyond a particular political economic theory's purview. Putting aside the moral implications of specialization, theorists would agree overall that increased territoriality has contributed to the minute division of labor under capitalism, making work impersonal, and circumscribing hierarchies of knowledge and responsibility. In addition neo-Smithians and neo-Marxians can agree that there exist for both capitalist and non-capitalist societies dynamic interrelationships among territoriality, technology, and such organizational characteristics as centralization, hierarchy, and span of control. Outlines of these interconnections were part of the internal dynamics of the theory of territoriality and were illustrated in Figures 2.1 and 2.2. Neo-Smithians and neo-Marxians can build on these particular dynamics, but other social theorists who have focussed on the narrower questions of organizational structure *per se* can also help specify them. As was mentioned in the discussion of territorial theory and illustrated in Chapter 4, on the Church, these interrelationships become more precise when the theory of territoriality is combined especially with offshoots of Weberian models of organization. We will turn now and examine territoriality in several contemporary institutions, focussing on the internal dynamics between territoriality and organizational structure while bearing in mind the neo-Smithian and neo-Marxian interpretations.

Territorial dynamics in contemporary settings

Chapter 2 pointed out that many of territoriality's potential effects are in opposition to one another. The advantage of not having to disclose what it is that is being controlled can also mean that the person controlling may not know what he has under control, and not defining by kind what is being controlled can be a cause of inefficient mismatches. Territorially subdividing knowledge and responsibility can make an organization more efficient, up to a point, by, for example, requiring fewer supervisors per supervisees – in other words by increasing the organization's span of control. But the same group of tendencies can lead to disorganization, to too much segmentation, and alienation (see Figure 2.2). Some of the general interconnections between territoriality and organizational structure have been illustrated through examples from history. The task now is to make them more precise and applicable to particular contemporary organizations. This means extending the internal logic of the theory to develop relationships that are suitable to modern organizations.

As noted in Chapter 2, a fruitful avenue for exploring the internal dynamics of the theory in a contemporary context would be first to focus on the hypothesized interrelationships among territoriality (*t*), span of control (*sp*) (which is the ratio of supervisor to supervisees), and geographical variability (*gv*). The interrelationship between the first two was discussed in Chapter 2 in the context of the combination (*d*) 'efficient supervision/span of control,' and leads directly to the hypothesis that, with everything else being equal (*ceteris paribus* can be assumed as well for the other relationships that follow), as territoriality (*t*) increases, span of control (*sp*) increases up to a point. Geographical variability (*gv*) and its relationship to territoriality (*t*) was discussed in Chapter 2 in the context of the third territorial effect (efficient enforcement of access) and leads to the connection of geographical variability (*gv*) of territoriality with geographic/temporal variability of the things to be controlled.

In addition, the theory's general proposition that territoriality is a strategy for establishing differential access points to the obvious connection between the utility of this strategy and the availability of other, non-territorial means of contact (*i*) which in turn depends in part on available technology. For example, if computers can substitute for teachers – an example of (*i*) – then there may be no need for students to meet together within the territory of the classroom. Two other variables linked to territoriality – level of individual skill and level of complexity of the task – are implicitly part of the theory, especially in discussions of work hierarchies. Skill and complexity are often closely related, and to simplify matters they can be combined in a single measure of complexity (*com*).

Summarizing, then, we have the following directly stipulated interrelationships, *ceteris paribus*:

1. as *t* increases, *sp* increases;
2. as *gv* increases, *t* decreases;
3. as *gv* increases, *i* increases;
4. as *i* increases, *t* decreases;
5. as *i* increases, *sp* increases.

Among the corollaries are *ceteris paribus*:

6. as *com* increases, *sp* decreases;
7. as *com* increases, *gv* increases;
8. as *com* increases, *i* increases;
9. as *com* increases, *t* decreases.

For convenience the complete set of hypothesized bivariate relationships can be placed on the following summary matrix.

	t	*gv*	*sp*	*com*	*i*
t		−	+	−	−
gv	−		−	+	+
sp	+	−		−	+
com	−	+	−		+
i	−	+	+	+	

Each cell contains the hypothesized direction of change between pairs of variables, holding everything else constant.

To test these interrelationships means that organizations must be found for which data on two or more of the five variables exist and for which the values of these variables change while virtually everything else is held constant. It is the second part that makes the selection so difficult from among even contemporary cases, let alone historical ones. Nevertheless, there appear to be data for several organizations approaching these conditions. One of these is the military.

Up to now our discussion of micro territoriality has involved indoor work and it may seem peculiar to shift to an 'outdoor' example, such as the armed forces. But the military nevertheless provides a work environment. Its structure is clearly hierarchical and authoritarian. Its purpose is explicit and its form is the same much the world over. This clarity and uniformity allows us to focus primarily on the hypothesized interrelationships of territoriality and organizational structure without having to wonder about who is controlling whom and for what purposes.

From among the services, the Army offers the clearest cases of degrees of territoriality, and official Army manuals describing military organizations have been available for decades.[88] From these and other sources[89] can be inferred the degree to which a military unit's tasks are territorial (*t*), along with measures for the unit's spans of control (*sp*), forms of indirect contact (*i*), complexity/skills (*com*) and the geographical and temporal variability of its tasks (*gv*).

For example, the lowest levels of combat units within the Infantry, Rangers, and Green Berets or Special Forces differ from one another in the degrees of territoriality of their missions, in their spans of control, in their degree of indirect contact, and in the complexity of their tasks. Their degrees of territoriality (*t*) can be inferred from descriptions of their missions. So too can the geographical variability (*gv*) of their tasks. Their spans of control (*sp*) are simply the numbers of officers per men at each level in the hierarchy; channels of communication (*i*) are measured by the amount of communications equipment assigned to each unit; and skills and complexity of task (*com*) are indicated by the ranks of the men and officers.

Located along a continuum that extends from very territorial to slightly or non-territorial, we have respectively the Infantry platoon, which has as one of its principal objectives the 'maintenance and security of terrain;' the airborne Rangers, whose principal objective is scouting and whose only territorial function is 'the securing of target objectives' (a smaller scale territory to be held for a brief period of time); and the Special Forces or Green Berets, whose objective is unconventional warfare. This means 'operations which include but are not limited to guerrilla warfare, evasion and escape, subversion and sabotage, conducted during periods of peace and war in hostile or politically sensitive territory.' Holding territory is not one of its stated missions.[90]

In other words, the objectives of the Infantry are most stationary and can be approached territorially, and the objectives of the Green Berets are least stationary and least territorial. (There may be other non-territorial geographical differences in their objectives, but what they are is not readily apparent.) From our theory we would expect that, everything else being equal, the more territorial the objective, either or both the greater will be the span of control and the less will be the need for communication and varied skills. Conversely, if the objective becomes less territorial and the task more complex and geographically varied, then there will be either or both an increase in communication and skill and a decrease in span of control. (We cannot predict which one or in what proportions.)

The Army's 1970 description of its own organization conforms to our expectations. Selected units have the following compositions.

The most territorial, the Infantry battalion, has:

at the level of rifle company,
  6 officers (of whom one is a captain), and 165 enlisted men;

at the level of rifle platoon,
1 officer (a lieutenant) and 43 enlisted men;
and at the level of rifle squad,
0 officers and 10 enlisted men with the following ranks:
1 staff sergeant, 2 sergeants, 4 specialists 4, and 3 privates first class;
also at the squad level are two pieces of radio equipment.

The intermediate group territorially, the airborne Rangers, has:

at the company level,
8 officers (of whom 2 are captains), and 208 enlisted men;
at the level of platoon,
3 officers (all lieutenants) and 129 enlisted men;
and at the patrol level,
0 officers and 5 enlisted men with the following ranks:
1 staff sergeant, 1 sergeant, 2 specialists 4, and 1 private first class;
also each patrol has three pieces of radio equipment.

For the Green Berets (who are the least territorial), the units are called detachments and are usually of three types: 'A,' 'B,' and 'C.' 'A' detachment is the basic operational unit and has 2 officers (a captain and a lieutenant) and 10 enlisted men, all of whom are sergeants of various ranks. 'B' detachment is a small command unit and has 2 officers (a major and a captain) and 3 enlisted men, all of various ranks of sergeant. 'C' detachment has 7 officers (1 lieutenant colonel, 2 majors, and 4 captains) and 15 enlisted men, all of various ranks of sergeant. There is no list of standard equipment for these units. Their needs vary per mission and in general they will have little communication equipment, for the preferred form of contact among themselves is face to face.

From these data, we can see that the Army, although reluctantly, conforms to our prediction. It alters its requirements for span of control, forms of communication, and skills of its men as the degree of territoriality of the mission changes. The Infantry units have a high span of control, the lowest number of higher-ranking officers per troops and the fewest number of officers per privates first class. The Ranger Company has a greater number of sergeants per privates first class at the platoon level and one more captain per company than the Infantry. Also, Rangers are divided into smaller units – patrols – and have more radio equipment. (If they did not, one would expect a lower span of control.) The Special Forces or Green Berets have a remarkably higher proportion of officers to enlisted men, and all the enlisted men have the rank of sergeant or above. (If there were more radio equipment per person, one might expect a slightly greater span of control.)

Examination of army organizations in other countries, and in the past (comparing, for instance, the territorial objectives and internal organizations of Cavalry and Infantry units), would likely bear out the hypothesized relationship. The more such studies are undertaken, the more precise can be the calibration of the interrelationships and their ranges.

This evidence was drawn from the organization's published rules. Rules may not be followed to the letter in practice but they do influence behavior. These manuals are distributed to troops and are used in strategic calculations. As anyone even slightly acquainted with Army life can attest to, the Army follows the 'manual' and the rules are as uniform and inflexible as can be.

Evidence for some of these interrelationships in other types of work places can be reconstructed from the multitude of existing studies on factory and work environments. These studies are based on actual behavior rather than on rules, although much of behavior at work is minutely stipulated by rules. Many of these studies were not intended to be about territoriality, nor do they address the five variables exactly as we have. Yet their data can be re-examined to indicate territoriality's effects. For example, one well-known comprehensive study of industrial workers measures the following characteristics of tasks:[91] Object Variety – the number of parts, tools, and controls to be manipulated; Motor Variety – variety in prescribed work place, variety in physical location of work, variety of prescribed physical operations of work; Autonomy – amount of worker latitude in selection of work methods, in selection of work pace, in accepting or rejecting the quality of incoming materials, in serving outside services; Required Interaction – number of people required to interact with at least every two hours, amount of time spent on required interactions; Optional Interaction on the Job – number of people available for interaction in working area, amount of time available for interaction while working; Optional Interaction off the Job – amount of time the worker is free to choose to leave the work area without reprimand; Knowledge and Skill – amount of time required to learn to perform job proficiency; Responsibility – degree of remedial action required to correct routine job problems.

In terms of the variables of the theory, territoriality (*t*) can be found in the measures of autonomy and optional interactions on and off the job. Geographical complexity (*gv*) is indicated by object and motor variety, and task complexity/skill (*com*) is indicated by knowledge/skill and responsibility. As we would expect from the theory, the data show that the indices we would use for territoriality are inversely related to those for complexity/skill and geographical variability. It is interesting to note that the study reveals also that surrogates for low territoriality and high complexity and geographical variability are positively related to job satisfaction.[92] This may not be surprising given what we know about minutely subdivided labor, territorial control, and the historical resistance of workers to factory discipline.[93]

The specific interrelationships among these variables can hold only when everything remains the same. But this happens rarely even in the laboratory. Work occurs in an ever changing social context and the attitudes of workers themselves are variable. Modern work requires a manageable workforce, to use Foucault's terms, of 'docile bodies.' Recently workers in the industrialized West have appeared less tractable. The 1960s and 1970s marked a period in the United States especially of severe cases of worker absenteeism, of low productivity, and of industrial sabotage.[94] These problems were part of a more general social discontent.

Ignoring the larger social context and focussing only on work, some have placed the burden for low worker productivity on the shoulders of the workers themselves. They have had it too easy. Unions have made jobs too cushiony. What is needed is greater discipline. Unions should be weakened, wages lowered, or else industry should pack up and leave to find cheaper and more pliable labor pools. In short the solution is even more industrial discipline. Others in management lay the blame at excessive (and often poor) management. Long-range planning and technology have been sacrificed for short-run profits. Automation and minute specialization have made jobs monotonous and work appear senseless. Jobs should be redesigned and made more interesting, and workers should be given greater responsibility.

Proposals for making work more attractive have ranged from essentially cosmetic changes in job description, to actually increasing the complexity of jobs, to having workers become involved in management.[95] Of those that increase the (real or apparent) complexity and responsibility of tasks, most imply, if not explicitly built into the new description, a less strictly territorial control over work. This is of course what territorial logic would suggest. Just as tasks were made simple by using territoriality to thin out the environment, relaxing territoriality would help to 'thicken' the environment and make tasks more complex. As the theory expects, and work in industrial management demonstrates, for specific tasks, job complexity, discretionary movement in space and time, and job satisfaction are often interrelated. Territorial restrictions would certainly have to be lifted to some degree if workers were to participate in management. If they are to help run the company, they must also have access to most areas of the work place. Even the more modest attempts at increasing worker responsibility often require giving workers correspondingly greater geographic freedom.

One of the pioneering efforts at increasing job complexity and worker responsibility was undertaken in the Swedish automobile industry. In the mid-1970s, Volvo and Saab gave workers more complex tasks by altering the assembly line to include 'work teams' and 'buffers.' A work team for automative body assembly is composed of seven workers;

> six work in pairs while the seventh serves as a coordinator, seeing that materials arrive and stepping in when someone is temporarily absent. In some ways the coordinator has the function of the foreman, but there is one important difference:

the position is rotated among members of the group on a weekly basis. The team not only builds auto bodies, it has taken many of the functions heretofore reserved for skilled craftsmen and white collar employees. It does most of the maintenance of its machinery, most quality control and, in consultations with management, hires new members and controls a . . . budget for new equipment. It can also, within limits, hire temporary replacements for ill members.[96]

The buffer is a means of storing work so that the

next piece . . . does not arrive at a mechanical tempo dictated by the line but is removed from the buffer stock; [the] . . . team's production goes into the buffer between it and the subsequent group. The consequence of this system is that the autonomy of the group is radically increased. The group controls its own production tempo. Faster production to fill the buffer permits longer breaks, often up to one and a half hours. The buffer system allows each group to control its own work process more or less as it likes. The limits are defined by the size of the buffer, but, within them, the group sets its own tempo, determines its schedule and if the outgoing buffer is completely filled, can simply take off for an hour or more.[97]

Work teams and buffer zones are not without problems. 'The union has had difficulty persuading groups to hire women and older workers who might cut group efficiency . . . and personality problems have developed.' And the buffer system is expensive to maintain, 'because it requires both space for the buffers between groups and ties up capital in stock.'[98] However, it has had its benefits. It has meant greater flexibility and lower costs for management, more interesting work and greater control of the work process for laborers.[99] These changes have lowered absenteeism and turnover.[100]

Management has experimented with other less hierarchical and centralized systems. One example is called matrix organization.[101] This system places individuals in overlapping sets of decision-making groups. Each individual belongs to more than one group and whenever possible decisions are made by consensus. An interesting attempt at introducing matrix organization occurred within one of the most traditionally rigidly hierarchical and territorial work environments: the ship. Indeed, life and work on board a ship is enormously territorially partitioned. Space is apportioned according to rank. The history of ships may provide the oldest continuous example of the relationships between territoriality and hierarchy. As for modern experiments, a Norwegian merchant shipping company attempted to 'democratize' the ship; to have decisions made non-hierarchically, through organizational matrices and consensus. It became apparent that an increase in flexibility meant a decrease in territoriality. Therefore, plans for new ship organization and design included 'equalizing of living conditions; common dayroom for all on board, including a library, bar, etc.'[102]

Examples of increasing flexibility of tasks and decreasing territoriality can be found in the work environment of the school. Elementary schools have attempted to create less rigidly structured classroom atmospheres. In place of the traditional 'closed' classrooms in which the students are anchored to their desks (as are workers to work stations in a factory), receiving a rigid schedule of tasks from the teacher who is usually stationed in the front of the room supervising the class, schools have experimented with an open classroom in which the children are permitted to move from place to place within the room to select their own subject and work at their own pace. Open classrooms often contain more than one grade per room, thus permitting the children a greater range of opportunities from which they can choose.[103] The assumption is that allowing the child more freedom to select what he wants to learn and when he wants to learn it will increase his overall interest in learning. (Whether or not this actually decreases span of control has not been determined.)

Work and territoriality do not form a closed system. This makes it very difficult to have confidence that the interrelationships which hold in one context will hold in another. So much of the calibration depends on broader social concerns and how the work environment is perceived. Research on office design illustrates this complexity. In a comprehensive review of the literature Oldham and Brass pointed out that two opposite effects of open office design can be expected.[104] On the one hand, what the authors termed the social interaction approach would lead us to expect that an open office plan would encourage interaction, that greater interaction increases friendships, and these would increase job performance. On the other, what the authors termed the socio-technical approach would argue that a lack of clear physical barriers, such as walls, will reduce the clarity of tasks and the sense of individual autonomy and thus reduce overall work effectiveness. Each uses different parts of the effects of territoriality. The first emphasizes (as do those who call for less division of labor) the job enrichment that comes from less territoriality, and the second (as do those who call for greater division of labor), the clarity of definition that comes from greater territoriality. The same authors attempted to determine which effects actually operated in a particular case.

They examined the attitudes of workers in a medium sized newspaper before and after they were moved from a closed to an open office structure. The move occurred in part because management expected an open office plan to increase the sense of group cohesiveness – to create a 'family like' atmosphere.[105] The results of their extensive attitudinal survey suggest that in this case the workers were less satisfied after the move than before, and primarily because of the lack of privacy, the lack of discreteness of their tasks and of their sub-groups. They found the open office plan noisier, describing it like 'Grand Central Station,' or like 'living in a fishbowl.'[106]

This particular case suggests that the tipping point had been reached between decreasing territoriality, increasing job variability, and job satisfaction. But why was it reached? It is difficult to answer this despite the study's elaborate experimental design. The new office was indeed excessively open and noisy. A few partitions and a lessening of noise may have made the workers feel that the move had been an improvement. But perhaps a more basic clue can be found in management's reasons and procedures for the new undertaking. The change was prompted originally by a desire on the part of management to increase the workers' sense of corporateness, but apparently the design was made without the consultation and assistance of the employees. Perhaps management's motivation, and the workers' feeling of being in a fishbowl, point out that from the beginning there were serious worker management problems which would have been aggravated by any change initiated by management.

Developments in new types of communication can be expected to affect territorial and non-territorial spatial organization. We have seen that the Catholic Church has relaxed some of its territorial hold over its parishioners in light of their increased geographical mobility and that television has provided some evangelical ministers with national congregations. New means of 'keeping in contact' have also lessened territorial control of prisoners. Convicts can be paroled earlier if they allow a small transmitter to be worn around a wrist or ankle so that the police would be able to monitor their whereabouts.[107] These are cases of technology changing access and of using the most efficient spatial strategy (in this case a non-territorial one) to maintain contact.

Coupling territoriality's logic with organizational theory has provided specific hypotheses about territory, job complexity, hierarchy, and technology. The openness of the work environment makes it difficult to fine tune even these specific relationships for a particular case, and next to impossible to generalize about them and make predictions. The most that can be expected is to be aware of possibilities, which is the case for social theory in general. A good social theory helps make sense of past and present events and points out what could reasonably be expected if certain conditions obtain.

If the interrelationships between territoriality and changing managerial strategies and technologies are important, but difficult to predict, then what of changes in social relations? How would they affect territorial organization and vice versa? Would democratization of the work place reduce its need for territoriality? Would socialism and communism reduce territorial control in the work place and in the political realm? Or is territorial organization of some kind or other essential to any complex and technologically advanced society?

If we encounter difficulties comparing territorial effects in one contemporary work place with another, it would be foolhardy to expect precise answers to such broad, yet important, questions. All that can be suggested is that territoriality, though in different forms and with different effects, seems to be an ubiquitious element of organization in all but the most primitive societies, although when it has been used there, it is as a device to control resources between, if not among peoples. Furthermore, it is clear that territoriality alone cannot alter social relations to the point of changing the complexion of an entire society, but it can, through its own internal dynamics, set in motion heretofore unforeseen, and often undesirable, social consequences. This was true with ancient civilizations. It was true with the Catholic Church. It was the case with the American territorial system and the work place. By all accounts it is true in socialist countries and one can expect it will be true in attempts at establishing more utopian communal organizations. Territoriality's effects are multiple, important, and must be reckoned with.

7

Conclusion: society, territory, and space

Space and time are fundamental components of human experience. They are not merely naively given facets of geographic reality, but are transformed by, and affect, people and their relationships to one another. Territoriality, as the basic geographic expression of influence and power, provides an essential link between society, space, and time. Territoriality is the backcloth of geographical context – it is the device through which people construct and maintain spatial organizations. For humans, territoriality is not an instinct or drive, but rather a complex strategy to affect, influence, and control access to people, things, and relationships. Its geographical alternative is non-territorial spatial behavior. Focussing on the latter has led geography and social science to emphasize the effects on human behavior of such metrical properties of space as distance. Unfortunately, this focus has been too constraining to permit development of a complex spatial logic. Adding a territorial component with a non-metrical emphasis to geographical analysis can help expand the logic of space, making it more flexible and realistic by imbedding it in social relations.

Territories are socially constructed forms of spatial relations and their effects depend on who is controlling whom and for what purposes. The task of the theory of territoriality is to disclose the possible effects of territoriality at levels that are both general enough to encompass its many forms, and yet specific enough to shed light on its particular instances. The multiple and complex interrelationships among the tendencies and combinations constitute the theory's internal dynamics. Some of these, like divide and conquer, are familiar. Most are not. Even the familiar ones are made clearer when their logical connections to other territorial effects are specified. We understand more of territoriality's role in dividing and conquering when we realize that territoriality can result in the joint employment of a specific group of tendencies, and that using these with slightly different emphases can help organizations become hierarchical and bureaucratic or can lead them to inefficiencies rather than help them to divide and conquer.

The theory's complex internal structure – its loops and tipping points – reveal territoriality's own internal dynamics. Which of the effects and interrelationships will actually be used, and what their import would be, depends on the social contexts of who is actually using territoriality and for what purposes. Specifying the context is not simply a matter of disclosing facts, for the evidence we marshal and the interpretations we give them depend on our ideas or models of who is in charge, from the personal level through to the societal. This book has explored only a few of the many possible connections among territoriality, theory, and context. Re-examining territoriality's link to these, as well as exploring its connections to others, offers a broad avenue for future research.

Conjoining territoriality as we did to a few broad models of social relations reveals that some of territoriality's effects can occur in practically any society. Others, though, are most often associated with particular political economic organizations. The rise of civilizations, and the rise of capitalism and modernity, are the two historical transitions which have seen the greatest changes in territoriality. In the first, the most important change was the use of territoriality to define and control people within a society as well as between societies; and in the second, was the use of territoriality to effect a sense of emptiable space, of impersonal relations, and of obscuring sources of power. Whereas territoriality underwent two major historical changes, social space and time appear to have undergone only one change in meaning of comparable proportions, and this occurred with the rise of capitalism and modernity. Territoriality, and the changing meaning of space and time, did not themselves occasion these political economic changes, but they did play fundamental roles in specifying the function and meanings of change.

The historical and global relationships among territory, space, time, and society are far too important and complex for us to have done more in a single volume than sketch their highlights. Even the focus on the role of territoriality in modern times was examined within only one cultural context. Doubtless, the three modern territorial facets – although critical to modernity everywhere – appear with different emphases in parts of the world other than the United States. In Europe, for instance, the development of an abstract sense of space did not occur as soon at the political territorial level as it did in the United States. And in Third World countries it still has not penetrated to as many levels as in North America. Indeed, the African continent, which, like the North American, was carved up abstractly and geometrically by colonial powers, retains, at the local level, much more of a sense of a social definition of territory, even though many of these tribal territorial areas were originally colonial demarcations that have since become instilled with social meanings. Similarly in Africa and elsewhere in the Third World, land holdings, whether personal or social, are not as unencumbered with historical and cultural meanings as they are in North America that they can appear as emptiable and fillable parcels. Although in these societies modern Western uses of territory have become more important, they are still deeply intermixed with complex patterns of pre-existing meanings and uses, and form different mixtures and intensities of old and new than are found in the North American example. Even Japan, a modern industrial society by any standard, uses these modern potentialities of territoriality with different intensities than does the United States. Comparing these mixtures is another reason to caution against associating territorial changes entirely with political economic ones. Just as culture, tradition, and history mediate economic change, they also mediate the way people and place are linked, the way people use territoriality, and the way they evaluate land.

Even in the North American context we can point to the persistence of pre-modern forms as well as to strategies about space and time that have been adopted to counter the prevailing experiences of modernity. And here too these attempts are stamped by culture. Americans have attempted to anchor themselves in space by creating historical landscapes in the forms of historical parks, national monuments, outdoor museums, historical preservation areas, and even pre-historical landscapes in wilderness areas. Americans have encouraged a love and allegiance to place in nationalism and patriotism, and they have sought for roots in neighborhood and small town America. Another antidote to viewing space as cold and abstract is the effort by geographers to introduce humanistic components to geographical analyses: to remind us that space is not experienced in everyday affairs only as an abstract emptiable framework on which events are contingently related. Rather space and place are filled with content and meaning. But modern, and especially American, society uses other and perhaps more pervasive means to place personal content in a space that is cold and abstract. Ironically these means are themselves part of the same processes that have helped impart these cold and impersonal qualities to geographical context.

We have seen that the development of an abstract metrical space went hand in hand with capitalism's need to increase production and consumption. An abstract space and time provide the economy with a powerful, practical, and easily manipulated framework to organize people and resources for a mass society. Also we have seen that this same spatial system makes it difficult to feel at home, to be rooted to place. At the same time that we decry the loss of personal ties and attempt to re-establish roots, the economic system, with its emphasis on consumption, holds out the promise of a solution to these paradoxes. It is through consuming products, we are told, especially by advertising, that we can once again be in the center of the world; that we can establish control over our own destinies. The act of consuming is supposed to create warm and meaningful contexts – both small and large, both ephemeral and substantial – and thereby it claims to address many of the problems consumption itself creates. Yet in fact consumption further exacerbates the problem because it is through consumption that the geographical and historical contexts are further fragmented, abstracted, and juxtaposed.

The economic order once again is undeniably of major importance. But its forms and effects depend on contexts. The theory of territoriality recognizes that differences exist in the degree to which modernity is manifested in different societies and that these differences may be due not only to cultural contexts but also to organizational dynamics. Organizational structures, like bureaucracies, may be relatively immune from all but major changes within political-economic systems, and some organizations may change far more than do their broader social contexts. There is, in other words, a dynamic to organizations which territoriality shares. This dynamic is seen within the Catholic Church; it can be seen within the bureaucracies of capitalist and socialist countries. In terms of social theory this means that in addition to the economic division of labor there exist broad social divisions having significant effects on who is controlling whom and for what purposes. And the theory of territoriality helps to specify the more likely effects that can occur within complex social organizations. Whatever the goals of a society may be – whether they are capitalistic, socialistic, or communistic – and whatever the geographical scale – whether local, national, or global – a society, simply as a complex organization, will need territoriality to coordinate efforts, specify responsibilities, and prevent people from getting in each other's way. And since territoriality of one sort or another will likely be employed, we must be aware that it possesses its own potentialities to affect and control, and that some of these may be contrary to the goals of the society. To plan without these considerations in mind would be to neglect significant determinants of social relations. Territoriality, as a component of power, is not only a means of creating and maintaining order, but is a device to create and maintain much of the geographic context through which we experience the world and give it meaning.

Notes

1. The meaning of territoriality

1 For Chippewa history see Edmund Danziger Jr, *The Chippewas of Lake Superior* (Norman, Oklahoma: University of Oklahoma Press, 1978); David Horr (ed.), *American Indian Ethnohistory: North Central and Northeastern Indians*, vols. 1–6 (New York: Garland Press, 1974); Harold Hickerson, *The Chippewa and Their Neighbors: A Study in Ethnohistory* (New York: Holt, Rinehart and Winston, 1970); and Edwin Higgins in collaboration with the Whitefish Lake Indian Reservation, no. 6, *Whitefish Lake Ojibway Memories* (Cobolt, Ontario: Highway Book Shop, 1982).

2 Harold Hickerson, *op. cit.*, n. 1, p. 9.

3 The North American aboriginal Indians made maps, often portraying vast areas. These maps contained physical features and the locations of Indian groups but did not indicate land holdings or territorial boundaries. Nor did the maps use a consistent scale or perspective. In many cases these Indian drawings provided Europeans with their only information about the New World and were therefore used as the basis of many European maps of the area.

4 Eleanor Leacock, 'Introduction,' in Frederick Engels, *The Origin of the Family, Private Property and the State* (New York: International Publishers, 1972), pp. 19–20. See also Diamond Jenness, 'The Ojibwa Indians of Parry Island, Their Social and Religious Life,' *National Museum of Canada Bulletin*, 78 (1954) (Anthropological Series no. 17).

5 Max Farrand, *The Legislations of Congress for the Government of the Organized Territories of the United States: 1798–1895* (Newark, N.J.: Wm A. Baker, Printer, 1896).

6 For a discussion of these two aspects of concepts in philosophy of science see May Brodbeck, 'Meaning and Action,' in May Brodbeck (ed.), *Readings in the Philosophy of the Social Sciences* (New York: Macmillan, 1968), pp. 58–78; and for their applications in geography see Robert Sack, 'A Concept of Physical Space in Geography,' *Geographical Analysis*, 5 (1973), pp. 16–34.

7 As is shown in the remainder of this chapter, definitions of territoriality are important. There are no other definitions exactly like this one. Most state simply that territory is the defense of area. There are one or two that seem close in

intention to ours. See for example Ryda Dyson-Hudson and E. Alden Smith, 'Human Territoriality: An Ecological Reassessment,' *American Anthropologist*, 80 (1975), pp. 21–41; and Edward Soja, *The Political Organization of Space* (Washington, D.C.: Association of American Geographers, Commission on College Geography, 1971). For a review of some of the many meanings of human territoriality see Torsten Malmberg, *Human Territoriality: Survey of Behavioral Territories in Man with Preliminary Analysis and Discussion of Meaning* (The Hague: Mouton, 1980); and Mark Wiljanen, 'A Critical Examination of a Concept of Human Territoriality with a Case Study in the History of American School Design' (Madison: University of Wisconsin, Department of Geography, Masters thesis, 1983).

8 'Ordinary places' glosses over many special geographical meanings of place, region, or area. The classic study on the geographical meanings of place and region is Richard Hartshorne, *The Nature of Geography: A Critical Survey of Current Thought in the Light of the Past* (Lancaster: Association of American Geographers, 1939, and reprinted with corrections, 1961). Ron Johnston (ed.), *The Dictionary of Human Geography* (Oxford: Basil Blackwell, 1981) has more recent definitions.

9 Note that Von Thunen's zones and central place hinterlands are not territories, except for Christaller's administrative areas. See Walter Christaller, *Central Places in Southern Germany*, trans. by Carlisle Baskin (Englewood Cliffs, N.J.: Prentice-Hall, 1966); and Brian Berry, *Geography of Market Centers and Retail Distribution* (Englewood Cliffs, N.J.: Prentice-Hall, 1967).

10 For a debate on the subject see Robert Ardrey, *The Territorial Imperative* (New York: Atheneum, 1966); M. F. Montagu (ed.), *Man and Aggression* (London: Oxford University Press, 1968); Torsten Malmberg, *op. cit.*, n. 7; J. Edney, 'Human Territoriality,' *Psychological Bulletin*, no. 81 (1974), pp. 959–75; and C. Fischer, 'The Myth of "Territoriality" in van den Berge's "Bringing the Beasts Back In",' *American Sociological Review*, 40 (1975), pp. 674–6. Our analysis will demonstrate that many of the effects and advantages of territoriality are applicable only to humans.

11 According to R. Taylor, 'Human Territoriality: A Review and a Model for Future Research,' *Cornell Journal of Social Relations*, 13, no. 2 (1978), pp. 125–51, territoriality is not meaningful as an autonomous concrete object, but rather must be understood in terms of its functions which vary across levels and forms of social organizations.

12 Jean Piaget and Barbel Inhelder, *The Child's Conception of Space*, (London: Routledge and Kegan Paul, 1956).

13 For a review of the literature on animal territoriality see C. Carpenter, 'Territoriality: A Review of Concepts and Problems,' in Anne Roe and George Simpson (eds.), *Behavior and Evolution* (New Haven: Yale University Press, 1958), pp. 224–50; and John Calhoun, 'The Role of Space in Animal Sociology,' *Journal of Social Issues*, 22, no. 4 (1966), pp. 46–58.

14 Consider the numerous ethnographic studies which describe behavior in space as territorial. For instance see E. Speck, *Family Hunting Territories and Social Life of Various Algonkian Bands of the Ottawa Valley* (Ottawa: *Geological Survey Memoir 70*, Anthropological Series no. 8, 1915); A. Carr-Saunders, *The Popula-*

tion Problem: A Study in Human Evolution (Oxford: Clarendon Press, 1922); Robert Lowie, *The Origins of the State* (New York: Russell and Russell, 1962); and A. Radcliffe-Brown, 'Preface,' in M. Fortes and E. Pritchard (eds.), *African Political Systems* (London: Oxford University Press, 1941). The lack of differentiation in the last two especially have led to mistaken interpretations of Morgan's thesis of a shift from a social to a territorial definition of social relations with the rise of civilizations. A book using territoriality in the title but which does not distinguish between territorial and non-territorial behavior in space is E. Gordon Erickson, *The Territorial Experience* (Austin: University of Texas Press, 1981).

15 See Cornelius B. Bakker and Marianne K. Bakker-Rabdue, *No Trespassing: Explorations in Human Territoriality* (San Francisco: Chandler and Sharp, 1973).

16 Examples are the concepts of externalities and spillovers in economics and public policy (Robert Bennett, *The Geography of Public Finance* (London: Methuen, 1980)). Political geography has been in the position of studying territories such as nation states, but it has not emphasized the logic of territoriality. Rather it has tended to reduce the territorial to the study of non-territorial spatial relations as when political geographers study the shapes of states, their relative locations, the locations of their capitals, and the distributions of resources within their boundaries. Recently some interest by political geographers in territoriality has surfaced, but it has not yet led to a systematic analysis of its role in power. See Stanley Brunn, *Geography and Politics in America* (New York: Harper and Row, 1974); and J. Gold, 'Territory and Human Spatial Behavior,' *Progress in Human Geography*, 6 (1982), pp. 44–67. Some useful suggestions are found in Kevin Cox, *Conflict, Power and Politics in the City: A Geographic View* (New York: McGraw-Hill, 1973), chapter 2.

17 Examples are found in Erving Goffman, *Asylums: Essays on the Social Situation of Mental Patients and Other Inmates* (Garden City, N.Y.: Anchor Books, 1961); Robert Sommer, *Personal Space* (Englewood Cliffs, N.J.: Prentice-Hall, 1969); H. Prochansky, W. Ittleson, and L. Rivlin (eds.), 'Freedom of Choice and Behavior in a Psychological Setting,' in *Environmental Psychology: Man and His Physical Setting* (New York: Holt, Rinehart and Winston, 1970), pp. 173–83.

18 See Irwin Altman and William Haythorn, 'The Ecology of Isolated Groups,' *Behavioral Science*, 12 (1967), pp. 169–82; and Eric Sundstrom and Irwin Altman, 'Field Study of Territorial Behavior and Dominance,' *Journal of Personality and Social Psychology*, 30 (1974), pp. 115–24.

19 The most notorious example is Robert Ardrey, *op. cit.*, n. 10. Even Torsten Malmberg, *op. cit.*, n. 7, uses instinct in his definition. 'Human behavioral territoriality is primarily a phenomenon of ethological ecology with an instinctive nucleus, manifested as more or less exclusive space, to which individuals or groups of human beings are bound emotionally and which, for the possible avoidance of others, are distinguished by means of limits, marks or other kinds of structurings with . . . display, movement or aggressiveness . . .' (pp. 10–11).

20 Irwin Altman, *The Environment and Social Behavior: Privacy, Personal Space, Territory and Crowding* (Monterey, California: Brooks/Cole Publishing Co., 1975), p. 107.

21 Territory is, according to A. Parr, in 'Environmental Design and Psychology,' *Landscape*, 14, no. 2 (1964), pp. 15–18, 'The space which a person . . . claims as

his or their own, and will defend'; according to J. Gold, *op. cit.*, n. 16, p. 49, it is 'any form of behavior displayed by individuals and groups seeking to establish, maintain, or defend specific bounded portions of space.' These particular references were brought to my attention by Julie Fisher.

22 I am referring here especially to Edward Soja, *op. cit.*, n. 7, and Ryda Dyson-Hudson and E. Alden Smith, *op. cit.*, n. 7.

23 Reviews of these theories of spatial analysis and approaches are found in Douglas Amedeo and Reginald Golledge, *An Introduction to Scientific Reasoning in Geography* (New York: John Wiley, 1975), especially chapters 7 and 8; Peter Haggett, Andrew Cliff, and Allen Frey, *Locational Analysis in Human Geography*, 2nd edn (London: Edward Arnold, 1977); and Richard Morrill, *The Spatial Organization of Society* (North Scituate, Mass.: Duxbury Press, 1974).

24 The spatial logic associated with distance has been called the 'distance decay assumption.' It is most extravagantly presented in William Bunge, *Theoretical Geography* (Studies in Geography, Series C, General and Mathematical Geography, No. 1, Lund: C. W. K. Gleerup, 1966). Among the many discussions of the difficulties with the use of distance as a variable in the analysis of human behavior are Gunnar Olsson, *Birds in Egg* (Ann Arbor: Department of Geography, University of Michigan, 1975), especially his analysis of the geographical inference problem, p. 465; and Robert Sack, *op. cit*, n. 6.

2. Theory

1 The attempts to isolate the effects of distance are part of the more general efforts to isolate the effects of spatial relations. The difficulty has been that space and its properties are part of all phenomena and yet are measurable independently of phenomena. A distance between two events can be measured precisely regardless of what these two events are and what stands between them. But the effect which that distance may have on the interconnections between these two events (excluding gravitational interactions) will be modified by the types of things that stand between, or more precisely by the media through which the interactions are to take place. If the phenomena refer to two people who want to have a normal conversation and are standing three feet apart, the medium will usually be the air between them that will carry the sound. If the air is not still, the distance will have to be shorter than three feet. In other words the effects of distance cannot be generalized independently of the specific medium even though we can measure distances independently of the medium. To attribute a significance to distance or to other spatial properties without regard to the substances or medium to which they refer is to over-generalize about the significance of space. This over-generalization has been called a non-relational concept of space: Robert Sack, *Conceptions of Space in Social Thought: A Geographic Perspective* (Minneapolis: University of Minnesota Press, 1980), pp. 80–1.

2 The medium affects the role of distance in physical processes with one major exception: the Newtonian equations for gravity. According to Newton, the gravitational force between two masses is directly proportional to their sizes and inversely proportional to the square of their distances. This inverse effect of distance is not modified by the presence or absence of intervening material.

Distance itself, absolutely and independently of matter, affects gravitational attraction. But this is the only example even in Newtonian physics in which distance has such an effect. And this system has since been reformulated by Relativity Theory to the effect that distance is a field of force and is not empty space. Some influential geographers and sociologists interested in spatial analysis unfortunately focussed on this one example to the exclusion of the more applicable cases and used it as a model for the effect of distance on human interactions. Their formulations were called the 'laws of social physics' in which the Newtonian equation of gravitational attraction formed the primary model. The basic model they presented is the distance interaction equation which states that the interaction between two population centers is directly related to their population sizes and inversely related to the distance between them. These and related attempts to isolate distance's effects without regard to the intervening substances treat space non-relationally. For examples of such attempts see William Bunge, *Theoretical Geography* (Studies in Geography, Series C, General and Mathematical Geography, No. 1, Lund: C. W. K. Gleerup, 1966); William Warntz, 'Geography of Prices and Spatial Interaction,' *Papers and Proceedings, Regional Science Association*, 3 (1957), pp. 118–29. Warntz states: 'space and time are to be recognized not just as cost-incurring external frictions, but rather as dimensions of the economic system and hence to be treated isomorphically in the rigid pattern of mathematical physics' (p. 128).

3 Robert Sack, 'A Concept of Physical Space in Geography,' *Geographical Analysis*, 5 (1973), pp. 16–34.

4 There are many discussions of possible definitions of what is benign and malevolent and how these definitions themselves have normative implications. For a review of the geographical literature see Anne Buttimer, *Values in Geography* (Washington, D.C.: Association of American Geographers, Commission on College Geography Resource Paper 24, 1974); and Bruce Mitchell and Dianne Draper, *Relevance and Ethics in Geography* (London: Longman, 1982).

5 This point has been made by Ryda Dyson-Hudson and E. Alden Smith, 'Human Territoriality: An Ecological Reassessment,' *American Anthropologist*, 80 (1975), pp. 21–41; and perhaps first by Marshall Sahlins, 'The Social Life of Monkeys, Apes and Primitive Man,' in J. N. Spuhler (ed.), *The Evolution of Man's Capacity for Culture* (Detroit: Wayne State University Press, 1959), p. 58. For a recent discussion of spatial strategies in hunting-gathering societies see, B. Winterhalder and E. Smith (eds.), *Hunter-Gatherer Foraging Strategies: Ethnographic and Archeological Analyses* (Chicago: University of Chicago Press, 1981).

6 This neutrality of territoriality has been identified by J. Edney as one of its most important effects. He states: 'the prime function of territoriality is precisely the thing that makes its true and most useful nature so elusive: to provide sufficient organization, structure, and predictability to be allowed to fade into a reliable background which the occupant does not have to concern himself with, but which is crucial for the subsequent development of more advanced behaviors and . . . efforts. The functions and benefits of territory would be drastically reduced if it

could not be taken for granted.' In 'Human Territories: Comment on Functional Properties,' *Environment and Behavior*, 8 (1976), p. 43.

7 W. Scott, *Rational, Natural and Open Systems* (Englewood Cliffs, N.J.: Prentice-Hall, 1981), p. 68.

8 Edward Soja, *The Political Organization of Space* (Washington, D.C.: Association of American Geographers, Commission on College Geography, 1971) reviews these and other anthropological perspectives.

9 James D. March (ed.), *Handbook of Organizations* (Chicago: Rand McNally and Co., 1965), *passim*.

10 A discussion of the modern facets involved in thinking of space as conceptually empty, and yet being able to describe it geometrically, is found in Robert Sack, *op. cit.*, n. 1.

11 Reification and displacement produce an effect similar to the one described as conflation of symbol and referent. See Robert Sack, *op. cit.*, n. 1, especially Chapter 6; and M. Mauss, *A General Theory of Magic* (London: Routledge and Kegan Paul, 1972), especially chapter 3.

12 See Chapter 6 in this book, and R. Edwards, *Contested Terrain* (New York: Basic Books, 1979); M. Katz, 'Our Origins of the Institutional State,' *Marxist Perspectives*, 4 (1978), pp. 6–22; and S. Marglin, 'What Do Bosses Do? The Origin and Functions of Hierarchy in Capitalist Production,' *Review of Radical Political Economics*, 6 (1974–5), pp. 36–60.

13 J. Pressman and D. Wildavsky, *Implementations: How Great Expectations in Washington are Dashed in Oakland* (Berkeley: University of California Press, 1973); G. Vernex, 'Overview of the Spatial Dimensions of the Federal Budget,' in N. Gluckman (ed.), *The Urban Impact of Federal Policies* (Baltimore: Johns Hopkins University Press, 1980), pp. 67–102.

14 S. Aguren, 'The Volvo Kalmar Plant' (Stockholm: The Rationalization Council, Swedish Information Services, 1976); and F. Emery and E. Thorsrud, *Form and Content in Industrial Democracy* (London: Tavistock, 1969).

15 Max Weber, *The Theory of Social and Economic Organization*, trans. by A. Hanerson and T. Parsons (New York: Oxford University Press, 1947).

16 K. Wittfogel, *Oriental Despotism: A Comparative Study of Total Power* (New Haven: Yale University Press, 1957).

17 Weber, *op. cit.*, n. 15, pp. 333–4.

18 D. Pugh and R. Payne, *Organizational Behavior in its Context: The Aston Programme III* (Westmead: Saxon House, 1977); and Joan Woodward (ed.), *Industrial Organization: Behaviour and Control* (London: Oxford University Press, 1970).

19 Robert Michels, *Political Parties* (Glencoe, Ill.: The Free Press, 1949).

20 Robert Merton, *Social Thought and Social Structure* (Glencoe, Ill.: The Free Press, 1957).

21 See Karl Marx's writings on Hegel, and A. Giddens, *Capitalism and Modern Social Theory* (Cambridge: Cambridge University Press, 1971), pp. 237–8, for a discussion of Marx and bureaucracy.

22 G. Konrad and I. Szelenyi, *The Intellectuals on the Road to Class Power*, trans. by A. Arate and R. Allen (New York: Harcourt Brace Jovanovich, 1979); Andras

Hegedus, *Socialism and Bureaucracy* (New York: St Martin's Press, 1976), especially chapter 1; Serge Mallet, *Bureaucracy and Technology in the Socialist Countries* (Nottingham: Spokesman Books, 1974); and Ernest Mandel, *On Bureaucracy: A Marxist Analysis* (London: I.M.G. Publications, 1973).

23 M. Dear, 'Theory of the Local State,' in A. Burnett and P. Taylor (eds.), *Political Studies From Spatial Perspectives* (New York: John Wiley, 1981, pp. 183–200; K. Newton, 'Conflict Avoidance and Conflict Suppression: The Case of Urban Politics in the United States,' in K. Cox (ed.), *Urbanization and Conflict in Market Societies* (Chicago: Maaroufa Press, 1978), pp. 76–93; and R. Walker and M. Heiman, 'The Quiet Revolution for Whom?', *Annals, Association of American Geographers*, 71 (1981), pp. 67–83.

24 Frederick Engels, *The Origin of the Family, Private Property and the State* (New York: International Publishers, 1972). As we will state in Chapter 3 more fully, the term primitive is not meant here as a pejorative but rather as a term referring to a form of social organization that existed before the rise of civilization and which persisted in some groups until the recent past. For the ideological component of the term primitive see Stanley Diamond, *In Search of the Primitive* (New Brunswick, N.J.: Transaction Books, 1974).

3. Historical models: territoriality, space, and time

1 There are no book length works devoted to the history of the relationships between political economy and geographical conceptions of space, though several works discuss key periods and issues. See for example, Ernst Cassirer, *The Philosophy of Symbolic Forms*, 3 vols. (New Haven: Yale University Press, 1953), *passim*; Anthony Giddens, *A Contemporary Critique of Historical Materialism* (Berkeley: University of California Press, 1981); George Gurvitch, *The Social Frameworks of Knowledge* (New York: Harper and Row, 1971); Stephen Kern, *The Culture of Time and Space 1880–1918* (Cambridge, Mass.: Harvard University Press, 1983); Donald Lowe, *The History of Bourgeois Perception* (Chicago: University of Chicago Press, 1982); Lewis Mumford, *Technics and Civilization* (New York: Harcourt Brace Jovanovich, 1934), especially pp. 9–28; Robert Sack, *Conceptions of Space in Social Thought: A Geographic Perspective* (Minneapolis: University of Minnesota Press, 1980); Pitirim Sorokin, *Socio-Cultural Causality, Space, Time* (New York: Russell and Russell, 1984); Yi-Fu Tuan, *Space and Place, The Perspective of Experience* (Minneapolis: University of Minnesota Press, 1977), and *Segmented Worlds and Self* (Minneapolis: University of Minnesota Press, 1982). For the most thorough Marxist interpretations of the uses of space by capitalism, see David Harvey, *The Limits to Capital* (Chicago: University of Chicago Press, 1982), and Neil Smith, *Uneven Development: Nature, Capital, and the Production of Space* (Oxford: Basil Blackwell, 1984).

2 See E. S. Deevey Jr, 'The Human Population,' *Scientific American*, 203 (1960), pp. 195–204, for population estimates.

3 For a discussion of the meaning of the term primitive see Stanley Diamond, *In Search of the Primitive* (New Brunswick, N.J.: Transaction Books, 1974).

4 As we noted in Chapter 2, the principal historical divisions upon which we will

concentrate are the primitive, pre-modern/pre-capitalist, and modern/capitalist. There are many other historical divisions and types of economies, as are there models of how they worked. The historical divisions in this book, as well as the models of their political economies, are ones that accommodate an emphasis on space, time, and territoriality. Needless to say they are sketches and draw from more general historical models. For the background to these divisions and modes see discussions of modes of production in Anthony Giddens, *op. cit.*, n. 1; Eric Hobsbawm (ed.), *Pre-Capitalist Economic Formations* (New York: International Books, 1975); Karl Polanyi, *The Great Transformation* (Boston: Beacon Press, 1944); and Eric Wolf, *Europe and the People Without History* (Berkeley: University of California Press, 1982).

5 What is being described here as a band is an abstraction containing a variety of types. For a discussion of patrilocal bands see A. Radcliffe-Brown, 'Preface,' in M. Fortes and E. Evans-Pritchard (eds.), *African Political Systems* (London: Oxford University Press, 1941); H. J. Heinz, 'Territoriality Among the Bushmen in General and the !Ko in Particular,' *Anthropos*, 67 (1972), pp. 405–16; and N. Peterson, 'Hunter-Gatherer Territoriality: The Perspective from Australia,' *American Anthropologist*, 77 (1975), pp. 53–68. For bilateral, fluid membership models see R. B. Lee and I. DeVore (eds.), *Man the Hunter* (Chicago: Aldine, 1968); and for an overall review see Elman Service, *Primitive Social Organization: An Evolutionary Perspective* (New York: Random House, 1962).

6 See Morton Fried, *The Evolution of Political Society* (New York: Random House, 1967), for a discussion of the 'stages' between band and chiefdom.

7 Note that in our society nationalism requires that the individual's primary allegiance be to the state, not the family or the local community.

8 Some of the interconnections among these factors are discussed in Marshall Sahlins, *Stone Age Economics* (London: Tavistock, 1974), chapter 2.

9 Examples of modesty about wealth and success are found in Stanley Diamond, *op. cit.*, n. 3.

10 The concepts of reciprocity, redistribution, and exchange to describe primitive, pre-modern, and modern economies respectively were originally formulated by Karl Polanyi, *op. cit.*, n. 4.

11 The idea that communities can ever be harmonious is a tragic vision according to Glenn Tinder, *Community: Reflections on a Tragic Ideal* (Baton Rouge: Louisiana University Press, 1981).

12 Even though there was disharmony an individual still felt enveloped by the community. Stanley Diamond, *op. cit.*, n. 3, p. 138.

13 T. Strehlow, *Aranda Traditions* (Carlton, Victoria: Melbourne University Press, 1947), pp. 30–1.

14 Frank Speck, 'Penobscot Tales and Religious Beliefs,' *Journal of American Folk-Lore*, 48 (1935), pp. 5–7; and H. Alexander, *The World's Rim* (Lincoln: Nebraska University Press, 1953), p. 89.

15 Ryda Dyson-Hudson and E. Alden Smith, 'Human Territoriality: An Ecological Reassessment,' *American Anthropologist*, 80 (1975), pp. 21–41.

16 Fr. Van Wing, *Etudes Bakongo*, 2nd edn (Bruges: Desclée de Brouwer, 1959), pp. 93–4.

17 See Marshall Sahlins' discussion of economic defendability in his 'The Social Life

of Monkeys, Apes, and Primitive Man,' in J. N. Sphuler (ed.), *The Evolution of Man's Capacity for Culture* (Detroit: Wayne State University Press, 1959), pp. 54–73.

18 Some of these issues seem to lie behind the argument made by Robert Lowie in his *The Origin of the State* (New York: Russell and Russell, 1962) when he attacked Morgan's view that primitive societies are based on social definitions of territory.

19 For a contrast between modern Western and traditional views see the discussions of space in Paul Bohannan, *Africa and Africans* (Garden City, N.Y.: Museum of Natural History, 1964).

20 This distinction between abstract space and space and places filled with content is made by Yi-Fu Tuan, in *Space and Place*, *op. cit.*, n. 1. Indeed the term space may be inappropriate in the setting of primitive society. Space is rather a series of places.

21 For a comparison of pre-modern and modern meaning of time see Lewis Mumford, *op. cit.*, n. 1; and Nigel Thrift, 'Spontaneity or Order: A Social History of Time Consciousness' (unpublished manuscript).

22 Alexander Marshack, *Roots of Civilization* (New York: McGraw-Hill, 1972).

23 See Edward Relph's discussion of inauthentic places in his *Place and Placelessness* (London: Pion Press, 1976).

24 Anthony Giddens, *op. cit.*, n. 1, uses the term class divided society.

25 For an overall review of theories concerning the transition to civilization see Elman Service, *Origins of the State and Civilization* (New York: W. W. Norton, 1975).

26 The example is drawn in part from Service's discussion on pp. 72–83 and the diagram is after Service, p. 77, in Elman Service, *ibid.*

27 The idea is similar to Service's discussion of Chiefly Ramage, *ibid.* p. 80.

28 For pre-capitalist forms such as this see Karl Marx, *Pre-Capitalist Economic Formations* (New York: International Publishers, 1965); and Barry Hindes and Paul Hirst, *Pre-Capitalist Modes of Production* (London: Routledge and Kegan Paul, 1975).

29 Robert McAdams, *The Evolution of Urban Society: Early Mesopotamia and Prehispanic Mexico* (Chicago: Aldine, 1966), finds several of these forms to pertain to Mesopotamia, see especially pp. 79–119. Yet they all emphasize territorial definitions of social relationships. About the importance of this change and its link to the rise of civilization McAdams says, 'Probably most would also tend to question Morgan's implicit assumption that the substitution of territorially defined communities for ethnically defined ones was both a necessary and a sufficient cause for the growth of the institution of private property. But there can be little dissent with the general shift he posited from ascriptively defined groupings of persons to politically organized units based on residence. And, substituting class stratification for property, I, for one, still share his conviction that it was the "mainspring" and "foundation" of political society' (quoting Morgan, p. 80).

30 See the discussion of how tribal demarcations became administrative areas of Ancient Egypt in Laurence Krader, *Formation of the State* (Englewood Cliffs, N.J.: Prentice-Hall, 1968), pp. 54–5. See also Frederick Engels' discussion of the

formation of Athenian jurisdictions in *The Origin of the Family, Private Property, and the State* (New York: International Publishers, 1972), pp. 176–9.

31 Accurate knowledge of the geography of large holdings and of political boundaries was nothing like it is today. A noble may know only that he owns an estate here and there, and it is the peasants on the estate who know where the particular boundaries of the property end. Political territories were often either defined by such estates or on an empirical level were more like frontier zones than sharply demarcated lines.

32 The order of territorial control in empires was first military, then administrative, then economic according to Michael Mann, 'State, Ancient and Modern,' *Archives Européennes de Sociologie*, 18, no. 2 (1977), pp. 262–98.

33 Joseph Needham and Wang Ling, *Science and Civilization in China* (Cambridge, Mass.: Harvard University Press, 1959), vol. 2, pp. 279–303. See also Paul Wheatley, *Pivot of the Four Quarters* (Chicago: Aldine, 1971), who discusses the relationship between symbolic geography and political integration. For similar use of symbolism in Andean society see, David Gow, 'Verticality and Andean Cosmology: Quadripartition, Opposition and Mediation,' *Actes du XLIIe Congrès International des Americanistes*, 4 (1978), pp. 194–211.

34 These points about Greek art are made by William Ivins, *Art and Geometry: A Study in Space Intuitions* (New York: Dover Publications, 1964).

35 Max Jammer, *Concepts of Space* (New York: Harper and Brothers, 1969), states that space was conceived by classical Greek philosophy and science at first as something 'inhomogeneous . . . and later as something anistropic . . . [and] these doctrines account for the failure of mathematics, especially geometry, to deal with space as a subject of scientific inquiry' (p. 23).

36 O. Neugebauer, *The Exact Sciences in Antiquity* (Providence, Rhode Island: Brown University Press, 1957), p. 225.

37 For a demographic analysis of the foundations of European capitalism see E. Jones, *The European Miracle* (Cambridge: Cambridge University Press, 1981); and for a political historical analysis see Perry Anderson, *Lineages of the Absolute State* (London: N.L.B., 1979).

38 This point is made most forcefully by Robert Brenner, 'The Origins of Capitalist Development: A Critique of Neo-Smithian Marxism,' *New Left Review*, 104 (1977), pp. 25–92. But it is not the only means by which people become dependent on the market. The New World farmers could have become subsistence farmers if it were not for the fact that they had to pay rent and mortgages in currency, not in kind. British colonialism in Africa forced African farmers into the market by levying taxes on land which had to be paid in money. This meant they had to produce for the market in order to pay the taxes.

39 Robert Brenner, *op. cit.*, n. 38, mentions the peasant farmers in parts of France as examples of those who could fall back on subsistence.

40 See Perry Anderson, *op. cit.*, n. 37, for a review.

41 For an analysis of the role of the state in capitalism, see Gordon Clark and Michael Dear, *State Apparatus: Structures and Language of Legitimacy* (Boston: Allen and Unwin, 1984), especially pp. 14–38; and Anthony Giddens, *op. cit.*, n. 1.

42 The history of the concept of progress is found in J. B. Bury, *The Idea of Progress*

(New York: Macmillan, 1932). For a similar discussion of "modern" see Crane Brinton, *The Shaping of the Modern Mind* (New York: Mentor Books, 1953), in which he states that 'Men have always lived in "modern" times but they have not always been as much impressed with the fact. Our own time, conventionally considered as beginning about 1500 A.D., is the first to coin so neat a term and apply it so consistently. Modern derives from a late Latin adverb meaning just now, and in English is found in its current sense, contrasted with ancient, as early as Elizabethan days. The awareness of a shaped newness, of a way of life different from that of ones forebears – and by 1700 awareness of a way of life felt by many to be much better than that of their forebears – this is in itself one of the clearest marks of our modern culture' (p. 19). See also his discussion of the idea of progress in the eighteenth century (p. 120).

43 Charles Trinkaus, 'Introduction: Renaissance and Discovery,' in F. Chiappelli (ed.), *First Images of America* (Berkeley: University of California Press, 1976), vol. 1, p. 8.

44 There are of course many characteristics to scientific concepts and to the language of science in general. What is meant here by clarity is described more fully in Robert Sack, *Conceptions of Space in Social Thought: A Geographic Perspective* (Minneapolis: University of Minnesota Press, 1980). Remoteness is also discussed there but in terms of the fact that scientific symbols are conventions, selected for their precision and for their significance and represented by symbolic vehicles that have nothing to do with the physical characteristics of the things being conceptualized. Thus a symbol x can stand for mass, y for volume and so on. This clarity of meanings and conventionalism of representation gives scientific symbols and concepts a psychological distance from what they represent. See Alfred Sohn-Rethel, *Intellectual and Manual Labour* (London: Macmillan, 1978) for a discussion of the interrelationships among capitalism, exchange, and abstract space and time.

45 Progress can be thought of as the rationalization of the material changes which capitalism brings about.

46 Nigel Thrift, *op. cit.*, n. 21, discusses the power that abstract time came to have. The difference between modern and pre-modern conceptions of time are discussed by Lewis Mumford, *op. cit.*, n. 1. Again, see Sohn-Rethel, *op. cit.*, n. 44, for an analysis of the relationships among capitalism and abstract space and time.

47 For analyses of large-scale medieval mapping see D. Chitton, 'Land Measurement in the 16th Century,' *Transactions, Newcomen Society*, 30 (1957–9), pp. 111–29; D. J. Price, 'Medieval Land Surveying and Topographical Maps,' *Geographical Journal*, 121 (1955), pp. 1–10; and H. Lyons, 'Ancient Surveying Instruments,' *Geographical Journal*, 69 (1927), pp. 137–43. Grids of some sort were used in a cartographic proposal of Roger Bacon, p. 11 of J. H. Parry, *The Age of Reconnaissance* (New York: World Publishing Co., 1963).

48 S. Edgerton, 'From Mental Matrix to Christian Empire,' in David Woodward (ed.), *Art in Cartography* (Chicago: University of Chicago Press, forthcoming).

49 *Ibid.*

50 J. H. Parry, *op. cit.*, n. 47, pp. 53–114.

51 There are, of course, exceptions as in the Inca policy of forest resettlement, and

the Mongols' attempts to remove the Chinese from the steppes, but these were either on a smaller scale or were unsuccessful when compared with North American attempts.

52 For a general discussion of the changing geography of Western cities, see J. Vance, *This Scene of Man* (New York: Harper College Press, 1977); and for the transition to the modern city see David Ward, *Cities and Immigrants* (New York: Oxford University Press, 1970).

53 Robin Evans, 'Figures, Doors, and Passages,' *Architectural Design*, 48, (1978), pp. 267–78, discusses the interior differentiation of the house.

54 See *ibid.* and Yi-Fu Tuan, *Segmented Worlds*, *op. cit.*, n. 1, for the relationships between architectural segmentation and the development of self-awareness. The language of architecture changed to reflect the concept of emptiable space. Until the modern period, architects designed buildings, now they also construct spaces and volumes. Compare the terms for space in Vitruvius or Alberti to the terms used in a modern book on architecture.

4. The Church

1 The terms visible, physical, or corporeal on the one hand, and invisible and incorporeal on the other, appear throughout discussions of the nature of the Church. See Sheldon Wolin, *Politics and Vision: Continuity and Innovation in Western Political Thought* (Boston: Little, Brown and Co., 1960).

2 Romans dedicated Basilicas and Jews dedicated their Temple. Reference to earliest Church consecrations is in Eusebius (His. Eccl. 10:3) wherein he described the dedication for the Basilica of Tyre, 314 A.D. The Catholic ritual for consecrating church buildings was not fully developed until the early Middle Ages.

3 Relics were important components in making places even more sacred. Many churches contained 'relics' of saints and martyrs. See R. W. Southern, *Western Society and the Church of the Middle Ages* (New York: Penguin, 1970), pp. 30–1.

4 The consecration of the church building and the subdivision of the church into gradations of sacred places was inspired by the Biblical descriptions of the Temple. As one Church historian in the seventeenth century put it: 'I hold that it is necessary to distinguish different degrees of holiness between the various parts of a church: and I have no doubt that the altar, for example, is more holy than the rest of the sanctuary; that the sanctuary is holier than the choir; that the choir is holier than the nave; and the nave is holier than the porch': J. G. Davies, *The Secular Use of Church Buildings* (London: S.C.M. Press Ltd, 1968), p. 97, quoting J. B. Thiers' study of church porches. Davies notes further that it is remarkable how close in thought this is to the Jewish Tractate Kelm of the Mishnah which states: 'There are ten degrees of holiness. The Land of Israel is holier than any other land . . . The walled cities are still more holy . . . Within the wall of Jerusalem is still more holy . . . The Court of Women is still more holy . . . The Court of the Israelites is still more holy . . . The Court of Priests is still more holy . . . Between the Porch and the Altar is still more holy, for none that has a blemish or whose hair is unloosed may enter there. The Sanctuary is still more

holy, for none may enter therein with hands and feet unwashed. The Holy of Holies is still more holy, for none may enter therein save only the High Priest on the Day of Atonement' (Davies, pp. 97–8).

5 There always have been Church officials who do not control territory, and there also have been other types of territories and ranks than the ones mentioned. In addition there is a non-territorial arm of control in the Church. For a discussion of Church authority see Stanley Chodorow, *Christian Political Theory and Church Politics in the Mid-12th Century* (Berkeley: University of California Press, 1972). For discussion of the structure of the diocese see *New Catholic Encyclopedia*, edited by Catholic University of America (New York: McGraw-Hill, 1967–74), vol. 4, under diocese.

6 For a review of the effects of Roman administrative structure on Church organization see A. H. M. Jones, *The Later Roman Empire* (Oxford: Basil Blackwell, 1964), vol. 2, pp. 873–985; for a discussion of a 'wage scale' within the Church, see pp. 904–10. For the general Roman structure see A. Sherwin-White, 'Roman Imperialism,' in J. Balsdon (ed.), *The Romans* (London: C. A. Watts and Co., Ltd, 1965), pp. 76–101. Even before the acceptance of Christianity by Rome, the *curses honorum* had begun. See E. Woodward, *Christianity and Nationalism in the Later Roman Empire* (London: Longman, Green and Co., 1916).

7 R. W. Southern, *op. cit.*, n. 3, especially pp. 56–90.

8 The historical data on the Church are vast and could offer a geographical gold mine of information for the patient scholar. But the data are scattered and incomplete. Categories change and it is difficult to develop even a superficial quantitative description of the system. A quantitative description can be had for certain segments of the Church and in certain regions and times, but these are often not generalizable. For an overview of Church history, see F. Mourett and N. Thompson, *A History of the Catholic Church*, 8 vols. (London: B. Herder Books, 1955).

9 There is a rich literature on the sociology of the Church. Concerning the differences between Church and other types of organizations see, for example, Thomas O. Dea, *The Sociology of Religion* (Englewood Cliffs, N.J.: Prentice-Hall, 1966). Concerning the organizational dynamics of churches see J. Benson and J. Dorsett, 'Toward a Theory of Religious Organization,' *Journal for the Scientific Study of Religion*, 10 (1971), pp. 138–51; J. Fichter, *Religion as an Occupation* (South Bend, Indiana: University of Notre Dame, 1961), especially pp. 263–4 and p. 271 when he discusses the impersonality of the Church structure; C. R. Hinings and B. D. Foster, 'The Organizational Structure of Churches: A Preliminary Model,' *Sociology*, 7–8 (1973–4), pp. 93–106; C. R. Ranson and H. Bryman, 'Churches as Organizations,' in D. S. Pugh and C. R. Hinings, *Organizational Structure: Extensions and Replications* (Westmead: Saxon House, 1976), pp. 102–14; and K. Thompson, *Bureaucracy and Church Reform* (London: Oxford University Press, 1970).

10 There are many lists of Apostles and the names do not always correspond, but for the three most familiar lists, see Luke 6, Mark 3, and Matthew 10.

11 See the references to the Temple in Peter Ackroyed, *Israel Under Babylon* (London: Oxford University Press, 1976); S. Zeithlin, *Studies in the Early History*

of Judaism (New York: Ktav Publishing House, 1973), vol. 1, pp. 143–75; and for a general discussion of Jewish attitudes towards Israel see Robert Cohn, *The Shape of Sacred Space* (Chico, California: Scholar's Press, 1981).

12 For a discussion see Peter Ackroyed, *op. cit.*, n. 11.

13 Ackroyed, *ibid.*, mentions an average of three pilgrims' feasts per year for which people could be expected to make the journey.

14 *Ibid.*, p. 26. By the time of Josiah attempts had been made to centralize all worship in Jerusalem and abolish all other sanctuaries, 2 Kings 22–3.

15 For the origins of the synagogue see K. Kohler, *The Origins of the Synagogue and the Church* (New York: Macmillan, 1929); and S. Sandmel, *Judaism and Christian Beginnings* (New York: Oxford University Press, 1978), especially pp. 131–2.

16 S. Sandmel, *op. cit.*, n. 15, pp. 131–4. There were other ranks within the Temple including the officers of the Temple, the treasurers and trustees, the controllers of the treasury, and the chief of the high priests.

17 Paul Johnson, *A History of Christianity* (New York: Atheneum, 1979), pp. 16–17; and M. Simon, *Jewish Sects at the Time of Jesus* (Philadelphia: Fortress Press, 1967), who says it is doubtful the Essenes had sacrifice.

18 Edwin Hatch, *Early Christian Churches* (London: Longman, 1981).

19 Wayne Meeks, *The First Urban Christians* (New Haven: Yale University Press, 1983), pp. 80–1.

20 See 1 Clement and the letters of Ignatius in Kirsopp Lake, *The Apostolic Fathers* (London: William Heinemann, 1912), vol. 1, especially 1 Clement XLII: 1–5, XLIV: 3; and Ignatius to the Smyrnaeans VII: 1–2, wherein he states: 'See that you all follow the bishop, as Jesus Christ follows the Father, and the presbytery as if it were the Apostles . . . Let no one do any of the things appertaining to the Church without the bishop . . . wherever the bishop appears let the congregation be present.' And in his letter to the Ephesians VII:1 he warns against heretics and about following false and itinerant prophets: 'For there are some who make a practice of carrying about the Name with wicked guile, and do certain other things unworthy of God; there you must shun as wild beasts, for they are ravening dogs, who bite secretly, and you must be upon your guard against them, for they are scarcely to be cured.' He continues in the Magnesians III:1, by equating the bishop with God. 'Now it becomes you . . . to render [the bishop] all respect according to the power of God the Father . . . As then the Lord was united to the Father and did nothing without him . . . so do you nothing without the bishop and the presbyters.'

21 C. J. Hefele, *A History of the Councils of the Church*, trans. by W. R. Clark, 5 vols. (Edinburgh: T. and T. Clark, 1895), are voluminous, and often contradictory. According to Hefele, vol. 1, pp. 67–75, the first general collection was by Merling in 1523. Several more were published in the sixteenth century. Hardouin's collection of 1722 and Mansi's of 1759 are widely referred to. There are also numerous regional collections. In the following, for the period up to the ninth century, we will be using the two major English compilations: those of C. J. Hefele and of E. H. Landon, *A Manual of Councils of the Holy Catholic Church*, 2 vols. (Edinburgh: John Grant, 1909). Even these two are not in complete agreement, although they do concur for most of the canons cited. Our wording of

the canons will often be paraphrases of these sources. In order to reduce the number of notes we have omitted references for each canon and its paraphrase. The canons are numbered and dated which allows the reader easily to consult the Landon and Hefele volumes. Since our intention in exploring the canons is to give the flavor and frequency of territorial issues, the precise dates and phrasings are not critical.

22 S. L. Greenslade, 'The Unit of Pastoral Care in the Early Church,' in G. J. Cumings (ed.), *Studies in Church History* (London: T. Nelson, 1965), vol. 2, pp. 102–18, especially pp. 106 and 112; and Ignatius to the Smyrnaeans VIII in Kirsopp Lake, *op. cit.*, n. 20.

23 See Paul's letters to the Corinthians, and Clement's first epistle to the Corinthians, and the letters of Ignatius to the Ephesians, Magnesians, Trallians, and Romans in Kirsopp Lake, *op. cit.*, n. 20.

24 S. L. Greenslade, *op. cit.*, n. 22, p. 106, states that 'I do not know how soon a bishop of any of these [Roman] cities thought of his pastoral care as bounded precisely by the limits of the *territorium* of his *polis*. Ignatius addresses Polycarp as bishop of the *ekklesia* of the Smyrnaeans, a congregational term. Did Polycarp regard himself as responsible for some thirty miles radius round the actual city of Smyrna? Perhaps he had not begun to think like that, but the time soon came when bishops did so envisage their pastoral charge.'

25 C. J. Hefele, *op. cit.*, n. 21, vol. 1.

26 An excellent review of early Christian attitudes towards the Holy Land is found in W. D. Davies, *The Gospel and the Land: Early Christianity and Jewish Territorial Doctrine* (Berkeley: University of California Press, 1974).

27 J. G. Davies, *op. cit.*, n. 4, p. 4, and quoting St Daniels.

28 J. G. Davies, *op. cit.*, n. 4, p. 10. There are, however, reports of Church ownership before this time. See A. H. M. Jones, *op. cit.*, n. 6, p. 895.

29 J. G. Davies, *op. cit.*, n. 4, p. 9.

30 *Ibid.*, p. 6.

31 C. J. Hefele, *op. cit.*, n. 21, vol. 1, p. 325.

32 A. H. M. Jones, *op. cit.*, n. 6, especially pp. 904–13, especially regarding the bureaucratic facets of the Church. As far as the territorial is concerned, Zeno had declared that there should be one bishop per city, although there were exceptions. For a discussion of the relationships between Church and civic boundaries see C. J. Hefele, *op. cit.*, n. 21, vol. 1, p. 382: 'The Church was not obliged in principle to conform itself to the territorial divisions of the states or of the provinces in establishing its own territorial divisions. If, however, it often accepted these civil divisions as models of its own, it was to facilitate the conduct of business, and to prevent any disruption of received customs. The Apostles often passed through the principal cities of one province for the purposes of preaching the gospel there before entering another, and often . . . they treated the faithful of that province as forming one community.' See also G. W. Addleshaw, *The Beginnings of the Parochial System* (London: St Anthony's Press n.d.) for a discussion of the territorial system in England.

33 J. G. Davies, *op. cit.*, n. 4, quotes one who resisted as saying 'What advantages accrues to him who reaches these celebrated places? He cannot imagine that our

Lord is living, in the body, there at the present day, but has gone away from us foreigners; or that the Holy Spirit is in abundance at Jerusalem, but unable to travel as far as us . . . You who fear the Lord, praise Him in the places where you are now. Change of place does not affect any drawing nearer to God, but wherever you may be, God will come to you' (p. 15).

34 E. H. Landon, *op. cit.*, n. 21, vol. 1, pp. 32–33.
35 C. J. Hefele, *op. cit.*, n. 21, vol. 2, p. 71.
36 E. H. Landon, *op. cit.*, n. 21, vol. 2, p. 126.
37 See discussion of the localization of the eucharist in the parish in *New Catholic Encyclopedia*, *op. cit.*, n. 5, vol. 10, pp. 1018–19.
38 Medieval authorities extended the symolism of the mass to the church building itself. See H. Taylor, *The Medieval Mind* (London: Macmillan and Co. Ltd, 1911), pp. 76–104.
39 Pope Galasius 492 A.D. is the turning point according to Gabriel Le Bras, 'The Sociology of the Church in the Early Middle Ages,' in S. L. Thrupp (ed.), *Early Medieval Society* (New York: Appleton-Century-Crofts, 1967), pp. 47–57.
40 F. Dvornick, *The Ecumenical Councils* (New York: Hawthorn Books, 1961), p. 48.
41 J. McNeill, 'The Feudalization of the Catholic Church,' in J. McNeill (ed.), *Environmental Factors in Christian History* (Chicago: University of Chicago Press, 1939). For an overview of Medieval Church organization see S. Baldwin, *Organization of Medieval Christianity* (New York: Henry Holt and Co., 1929). For the detailed case studies of Medieval Church administration see C. R. Cheney, *Episcopal Visitations of Monasteries in the 13th Century* (Manchester: Manchester University Press, 1931); and R. K. Rose, 'Priests and Patrons in the 14th Century Diocese of Carlisle, in *Studies in Church History* (Oxford: Basil Blackwell, 1979), vol. 16, pp. 207–18.
42 J. G. Davies, *op. cit.*, n. 4.
43 R. W. Southern, *op. cit.*, n. 3, p. 38.
44 Paraphrasing R. W. Southern, *ibid.*, p. 102.
45 *Ibid.*, p. 158.
46 Paul Johnson, *op. cit.*, n. 17, p. 273.
47 'Night and day I pondered until I saw the connection between the justice of God and the statement, "righteous shall live by his faith." Then I grasped that the justice of God is that righteousness by which through grace and sheer mercy God justifies us through faith.' Luther quoted in George Mosse, *The Reformation*, 3rd edn (New York: Holt, Rinehart and Winston, 1963), p. 25.
48 M. Luther, 'Church and Ministry I,' in E. W. Gritsched (ed.), *Works* (Philadelphia: Fortress Press, 1970), vol. 39, p. 219.
49 J. G. Davies, *op. cit.*, n. 4, discusses this on p. 120.
50 M. Luther, 'Treatise in Trade and Usury,' in 'The Christian Society II,' E. W. Gritsched, *op. cit.*, n. 48, vol. 45, p. 286.
51 M. Luther, 'Church and Ministry I,' in *ibid.*, vol. 39, p. 219.
52 M. Luther, 'Trade and Usury,' in 'The Christian Society II,' *ibid.*, vol. 45, p. 286.
53 M. Luther, 'Concerning the Ministry,' in 'Church and Ministry I,' *ibid.*, vol. 40, pp. 7–44.

54 G. Mosse, *op. cit.*, n. 47, pp. 29–31. See also Luther's 'Instructions for the Visitors of Parish Pastors in Electoral Saxony,' *ibid.*, vol. 40; and his 'Temporal Authority: To What Extent Should it be Obeyed,' *ibid.*, vol. 45.

55 M. Luther, 'Ordinance of a Common Chest,' *ibid.*, vol. 45, pp. 169–94.

56 E. G. Leonard, *A History of Protestantism*, trans. by J. Reid (London: Nelson, 1965), p. 114.

57 *The New Schaff-Herzog Encyclopedia of Religious Knowledge*, S. Jackson (ed.) (New York: Funk and Wagnalls Co., 1910).

58 See the *New Encyclopaedia Britannica*, 15th edn (Chicago, Encyclopaedia Britannica, 1974), vol. 3, p. 672, for a review of Calvin's Church government.

59 John Calvin, 'Articles Agreed upon by the Faculty of Sacred Theology of Paris with the Antidote,' in Henry Beveridge (trans.), John Calvin, *Tracts and Treatises* (Grand Rapids: Berdmans Publishing Co., 1958), vol. 1, p. 103.

60 John Calvin, 'Ordinances for the Supervision of the Churches in the Country,' in *ibid.*, pp. 223–47.

61 G. Mosse, *op. cit.*, n. 47, p. 71.

62 John Calvin, 'Draft Ecclesiastical Ordinances,' in John Dillenburger (ed.), *John Calvin Selections from His Writings* (Garden City, N.Y.: Anchor Books, 1971), p. 233.

63 W. Whittley (ed.), *The Works of John Smythe* (Cambridge: Cambridge University Press, 1915), p. 252.

64 The Puritan settlements of New England began with intentions of becoming theocracies. Their model of Church government by and large was an English one, and the parish originally played a large part in their organization and definition of community. Yet even they were to exhibit considerable flexibility regarding Church structure and hierarchy, sometimes emphasizing congregational and other times episcopal forms. See L. Bacon, *The Genesis of the New England Churches* (New York: Arno Press, 1972).

65 There is a vast literature on Protestant Church organization that could be examined fruitfully from a geographical perspective. For an overview, see F. E. Mayer, *The Religious Bodies of America* (St Louis, Mo: Concordia Publishing House, 1958); E. S. Mead, *Handbook of Denominations in the United States* (Nashville: Abingdon Press, 1970); and for particular constitutions see *The Book of Church Order of the Presbyterian Church in the United States* (Richmond, Va; The Board of Christian Education, 1963); E. S. Gaustad (ed.), *Baptists: The Bible, Church Order and the Churches* (New York: Arno Press, 1980); E. S. Gaustad (ed.), *Baptist Ecclesiology* (New York: Arno Press, 1980); N. B. Harmon, *The Organization of the Methodist Church: Historic Development and Present Working Structure* (Nashville: The Methodist Publishing House, 1962); M. Hoffman, *A Treatise on the Law of the Protestant Episcopal Church in the United States* (New York: Stamford and Swords, 1850); T. B. Neely, *The Bishops and the Supervisional System of the Methodist Episcopal Church* (Cincinnati: Jennings and Graham, 1912); J. M. Tuell, *The Organization of the United Methodist Church* (Nashville: Abingdon Press 1970); *United Church of Christ, History and Program* (Publication Board for Homeland Ministries, n.d.); E. White, *Annotated Constitutions and Canons for the Government of the Protestant*

Episcopal Church in the United States of America (Greenwich, Connecticut: The Seabury Press, 1954), vol. 2.

66 For modern Catholic theological views see Michael Novak, *Confessions of a Catholic* (San Francisco: Harper and Row, 1983).

67 See references to the Church's activities in South America in the *New Catholic Encyclopedia*, *op. cit.*, n. 5, vol. 9, pp. 945–6. The Treaty of Tordesillas may appear in retrospect to be an abstract clearing of space. The issue is discussed in Chapter 5 of this book.

5. The American territorial system

1 The British had essentially two types of authorized colonies. One was a grant to an individual in the form of a proprietorship or 'fiefdom' and the other was a charter to a corporation. There were unauthorized colonies such as Plymouth and Rhode Island. For discussions of the origins of New World grants and charters, see L. P. Kellog, 'The American Colonial Charter,' *Annual Report of the American Historical Association*, 1 (1903), pp. 185–341; A. Madden, '1066, 1776 and All That: The Relevance of English Medieval Experience of "Empire" to Later Constitutional Issues,' in J. Flint and G. Williams (eds.), *Perspectives of Empire* (London: Longman, 1973), pp. 9–26. For a discussion of Palatinates see, G. T. Lapsley, *The Country Palatine of Durham: A Study in Constitutional History* (New York: Longman, Green and Co., 1900); and J. Scammell, 'The Origins and Limitations of the Liberty of Durham,' *English Historical Review*, 80 (1966), pp. 449–73.

2 For a review of medieval cadasters and surveying techniques, see D. J. Price, 'Medieval Land Surveying and Topographical Maps,' *Geographical Journal*, 121 (1955), pp. 1–10.

3 T-O maps are discussed in Norman Thrower, *Maps and Man* (Englewood Cliffs, N.J.: Prentice-Hall, 1972), pp. 32–4. See also Lloyd A. Brown, *The Story of Maps* (New York: Bonanza Books, 1949), pp. 94–100.

4 Longitude and latitude took time to become implemented. It was not until the seventeenth century that movable time pieces made accurate measures of longitude possible. See J. H. Parry, *The Age of Reconnaissance* (New York: World Publishing Co., 1963), pp. 97–8.

5 Joseph Needham and Wang Ling, *Science and Civilization in China* (Cambridge, Mass.: Harvard University Press, 1959), vol. 2, pp. 279–303.

6 See James Carder, *Art Historical Problems of a Roman Land Surveying Manuscript* (New York: Garland, 1978); and P. Mackrindick, 'Roman Colonization and the Frontier Hypothesis,' in W. Wyman and C. Kroeber (eds.), *The Frontier in Perspective* (Madison: University of Wisconsin Press, 1957), pp. 3–20.

7 Edmund Burke's concept of virtual representation is found in his letter to Sir Hercules Langrishe, *Works of the Right Honorable Edmund Burke* (Boston: Little, Brown and Co., 1871), vol. 4, p. 293. A general analysis of the meanings of representation is found in H. F. Pitkin, *The Concept of Representation* (Berkeley: University of California Press, 1967). For reviews of systems of 'representation' before the modern era see J. R. Pole, *Political Representation in England and the*

Origins of the American Republic (New York: St Martins Press, 1966); D. Lutz, 'The Theory of Consent in Early State Constitutions,' *Publius*, 9 (1979), pp. 11–42; and George Sabine, *A History of Political Theory* (New York: Holt, Rinehart and Winston, 1961), especially pp. 206–301. R. MacIver, *The Modern State* (London: Oxford University Press, 1964), pp. 141–2, argues that representation as we know it had no real antecedents. The concept of organic community or *Gemeinschaft* (though not of organic representation) is discussed by F. Tonnies, *Community and Society* (East Lansing: Michigan State University Press, 1957). The distinction between *Gemeinschaft* and *Gesellschaft* can be traced from Durkheim back to Plato. *Gemeinschaft* or organic community is suggested by organic representation while the community of convenience, the *Gesellschaft*, is suggested by proportional representation.

8 Translations of the documents can be found in Henry Steele Commager (ed.), *Documents of American History* (New York: F. S. Crofts and Co., 1935), pp. 2–5.

9 See the discussion of Donation in L. Weckmann-Munoz, 'The Alexandrine Bulls of 1493,' in F. Chiappelli (ed.), *First Images of America* (Berkeley: University of California Press, 1976), vol. 1, pp. 201–9.

10 This, despite the fact that Columbus reports citing naked peoples and the Pope refers to them in the Treaty.

11 Henry Steele Commager, *op. cit.*, n. 8, p. 5. For a slightly different version see F. N. Thorpe (ed.), *The Federal and State Constitutions, Colonial Charters, and Other Organic Laws of the United States* (Washington, D.C.: Government Printing Office, 1909), vol. 1, pp. 45–6.

12 F. N. Thorpe, *op. cit.*, n. 11, vol. 1, p. 54.

13 John Smith quoted in Francis Jennings, *The Invasion of America: Indians, Colonialism, and the Cant of Conquest* (New York: W. W. Norton and Company, Inc., 1976), p. 78.

14 Purchase quoted in Francis Jennings, *ibid.*, p. 80.

15 *Ibid.*, p. 80. For more detail on attitude toward savages and their rights to land see Etienne Grisel, 'The Beginnings of International Law and General Public Law Doctrine,' in F. Chiappelli, *op. cit.*, n. 9, vol. 1, pp. 305–25; and W. Washburn, 'The Moral and Legal Justification for Dispossessing the Indians,' in J. Smith, *Seventeenth Century America* (Chapel Hill: University of North Carolina Press, 1959), pp. 15–32. Many found the attitude towards the 'savages' hypocritical. Montaigne said 'I find that there is nothing barbarous and savage in this nation, by anything that I can gather, excepting, that every one gives the title, barbarism, to everything that is not in use in his own country.' Michel Montaigne, *Essays of Montaigne* (London: Reeves and Turner, 1902), vol. 1, p. 241.

16 The Dutch West India Co. had been chartered as a commercial monopoly with semi-sovereign powers but without a grant of territory. The charter is found in A. J. F. Van Laer (ed.), *Van Renselaer Boweir Manuscripts* (Albany, N.Y.: N.Y. State Library, 90th Annual Report II, 1908), pp. 86–125.

17 N. Carmey, 'The Ideology of English Colonization: From Ireland to America,' *William and Mary Quarterly*, ser. xxx (1975), pp. 575–98.

18 A. Slavin, 'The American Principles From More to Locke,' in F. Chiappelli, *op. cit.*, n. 9, vol. 1, p. 139.

19 Roger Williams, paraphrased and quoted by William Cronon, *Changes in the*

Land: Indians, Colonists, and the Ecology of New England (New York: Hill and Wang, 1983), p. 57.

20 Francis Jennings, *op. cit.*, n. 13, p. 82, paraphrasing Winthrop.

21 William Cronon, *op. cit.*, n. 19, pp. 56–7, quoting Cotton. Later Massachusetts law did allow Indians to retain cultivated land.

22 Francis Jennings, *op. cit.*, n. 13, p. 60, quoting Marshall. Settling, clearing, planting, and in general introducing any type of capital investment has been, and still is, called an 'improvement' or 'development.' Housing contractors are called 'developers.'

23 F. N. Thorpe, *op. cit.*, n. 11, vol. 7, p. 3795.

24 S. Kingsley, *Records of the Virginia Company* (Library of Congress, 1906–35), vol. 3.

25 F. N. Thorpe, *op. cit.*, n. 11, vol. 1, pp. 69–70.

26 *Ibid.*, vol. 5, pp. 2754–5.

27 *Ibid.*, vol. 6, p. 3215.

28 *Ibid.*, vol. 3, p. 1632.

29 *Ibid.*, vol. 5, p. 2538.

30 *Ibid.*, vol. 5, p. 2574.

31 *Ibid.*, vol. 5, pp. 3036–40 (author's emphasis).

32 *Ibid.*, vol. 5, p. 2772.

33 Marshall Harris, *Origin of the Land Tenure System in the United States* (Ames: Iowa State College Press, 1953), pp. 37–8.

34 Harry Wright (ed.), *Indian Deeds of Hampden County, Massachusetts* (Springfield, Massachusetts, 1905), p. 11.

35 Amelia Ford, *Colonial Precedents of Our National Land System as It Existed in 1800* (Madison: University of Wisconsin, 1910), pp. 29–30, referring to Journals of the House of Representatives of Massachusetts Bay, 1715, p. 30.

36 See a description of Penn's grant in Marshall Harris, *op. cit.*, n. 33, p. 240. For a general discussion of early North American surveying and mapping see M. Eggleston, 'The Land System of the New England Colonies,' *The Johns Hopkins University Studies in Historical and Political Science Series*, 4, (Baltimore, 1886), p. 22; W. D. Pattison, *Beginnings of the American Rectangular Land Survey System, 1784–1800* (Chicago: Department of Geography, Resource Paper no. 50, 1957), especially p. 64; and N. J. Thrower, *Original Survey and Land Subdivisions: A Comparative Study of the Form and Effect of Contrasting Cadastral Surveys* (Chicago: University of Chicago Press, 1967).

37 Amelia Ford, *op. cit.*, n. 35, p. 30.

38 Marshall Harris, *op. cit.*, n. 33, pp. 338–9.

39 Harry Wright, *op. cit.*, n. 34, also referred to in William Cronon, *op. cit.*, n. 19, pp. 74–5.

40 Constantia Maxwell, *Irish History from Contemporary Sources (1509–1610)* (London: George Allen and Unwin, 1923), especially pp. 37, 279, and 281; and D. B. Quinn, *The Elizabethans and the Irish* (Ithaca, N.Y.: Cornell University Press, 1966).

41 See, for example, Captain Nicholas Pynnar's survey.

42 'By the King, a Project of the Division and Plantation of the Escheated Lands in Six Several Counties in Ulster, Namely Tyrone, Coleraine, Donegal, Ferman-

agh, Armagh, and Cavan,' in Walter Harris, *Hibernica; Or, Some Ancient Pieces Relating to Ireland* (Dublin: John Milliken, 1770), p. 105.
43 *Ibid.*, p. 107.
44 *Ibid.*, pp. 123–38, quotations pp. 135–6.
45 *Ibid.*, pp. 140–3.
46 *Ibid.*, pp. 140–3. Amelia Ford, *op. cit.*, n. 35, p. 2, mentions the relatively abstract layouts by Petty.
47 Calendar of Fiants, Edward VI, no. 724; Eliz. no. 474, contained in the 8th and 11th Reports of the Keeper of the Public Records of Ireland, and contained in House of Commons, 1878–9, vol. 39 and House of Commons, 1876, vol. 39.
48 This experiment is in a sense a non-metrical equivalent of the distance interaction concept in geography. See the discussion of spatial analysis in Chapter 1 of this book.
49 For a discussion of these matters in the Cambridge Platform, see *Creeds and Platforms of Congregationalism* (Boston: The Pilgrim Press, 1893 and 1960). See also John Cotton, *The Way of the Churches in New England* (London, 1645), pp. 109–10.
50 Robert Kingdon, 'Protestant Parishes in the Old World and New: The Cases of Geneva and Boston,' *Church History*, 48 (1979), pp. 290–304.
51 Robert Kelso, *The History of the Public Poor Relief in Massachusetts: 1620–1920* (Boston: Houghton Mifflin and Company, 1922), p. 36.
52 *Ibid.*, p. 38.
53 *Records of the Colony of New Plymouth in New England*, D. Pulsifer (ed.), vol. 7, *Laws: 1623–1682* (Boston: William White, 1861), p. 118.
54 Second Report of the Record Commissioners, City of Boston, Document 46–1881 (Boston Town Records, 1634–60); also Robert Kelso, *op. cit.*, n. 51.
55 Robert Kelso, *op. cit.*, n. 51, p. 41.
56 *Ibid.*, pp. 41–2.
57 *Ibid.*, p. 46.
58 *Ibid.*, p. 59.
59 D. T. Fiske, 'Discourse Relating to the Churches . . . of Essex North,' in *Contributions to the Ecclesiastical History of Essex County, Massachusetts* (prepared and published under the direction of the Essex North Association of Boston: Congregational Board of Publications, 1865), p. 263; *Massachusetts Historical Collections*, vol. 1, p. 216, and vol. 3, p. 78.
60 Resource is a relatively new term. The *O.E.D.* cites its first appearance in 1611.
61 For a discussion of the commercialization of land and its effects on colonial land use practices see William Cronon, *op. cit.*, n. 19, p. 75.
62 Thomas Hobbes, *Leviathan* (Chicago: Encyclopaedia Britannica, Great Books, 1952).
63 Jean Bodin, *Six Books of the Commonwealth* (Oxford: Basil Blackwell, 1955), p. 30.
64 John Milton, *Prose Work* (London: Everyman's Library, 1958), pp. 233–42; and David Hume, T. H. Greene, and T. H. Grose (eds.), *Essays Moral, Political, and Literary* (New York: Longman, Green and Co., 1898), vol. 1, pp. 482–93.
65 Jean-Jacques Rousseau, *Political Writings* (New York: Nelson, 1953).
66 Samuel Huntington, 'The Founding Fathers and the Division of Powers,' in

Arthur Maass (ed.), *Area and Power* (Illinois: The Free Press, 1959), pp. 150–205.

67 C. Kenyon (ed.), *The Antifederalists* (Indianapolis: The Bobbs-Merrill Co., 1966).

68 James Madison in M. Farrand (ed.), *The Records of the Federal Convention*, 3 vols. (New Haven: Yale University Press, 1911), vol. 2, p. 124. This follows the statement that class should be represented as is the case with geographical communities.

69 *Federalist*, no. 10.

70 James Madison, in G. Hunt (ed.), *Writings*, 9 vols. (New York: G. Putnam and Sons, 1900–1), vol. 5, p. 31. On factions and territory, see vol. 9, pp. 524–5; and in vol. 6, after discussing the virtues of the partition of powers among legislative, judicial and executive, he adds: 'the political system of the U.S. claims still higher praise. The power delegated by the people is first divided between the general government and the state governments; each of which is then subdivided into legislative, executive, and judiciary departments . . . [This] may prove the best legacy ever left by lawgivers to their country' (pp. 91–2). And in vol. 9 he asks: 'Into how many parts must Virginia be split before the semblance of such a condition [homogeneity of interests] could be found . . . in the smallest of the fragments, there would soon be added to previous sources of discord a manufacturing and an agricultural class . . .?' (p. 526).

71 Roy Honeywell, *The Educational Works of Thomas Jefferson* (Cambridge, Mass.: Harvard University Press, 1931), especially chapter 2.

72 Thomas Jefferson, *Writings*, 20 vols., Andrew Lipscomb (ed.), (Washington, D.C.: The Thomas Jefferson Memorial Association, 1903–4), vol. 2, p. 122, cited in Samuel Huntington, *op. cit.*, n. 66, p. 175.

73 Donald Lutz, *op. cit.*, n. 7. This of course did not mean that franchise was universal. See also J. Pole, *Political Representation in England and the Origins of the American Republic* (London: Macmillan, 1966), especially pp. 127 and 385; and G. Wood, *The Creation of the American Republic* (Chapel Hill: University of North Carolina Press, 1969), p. 172.

74 This may have been due to the fact that no one knew which states would ratify the Constitution.

75 Max Farrand, *The Legislations of Congress for the Government of the Organized Territories of the United States: 1789–1895* (Newark, N.J.: Wm A. Baker, 1896).

76 *Federalist*, no. 17. See also *Federalist*, no. 45.

77 Kenneth Winkle, 'The Politics of Community: Migration and Politics in Antibellum Ohio' (Madison: University of Wisconsin, Ph.D. thesis, 1984).

78 De Tocqueville comments frequently about the mobility of Americans.

79 It is more precise to say that they would agree if they were made aware of these effects, because many theorists ignore space and territory completely. They operate as though political economic activities occur in a geographical vacuum.

80 Adam Smith recognized the need for government to provide public goods. In this respect he seems more 'liberal' than are some American conservative politicians who leave virtually no room at all for the role of government.

81 The definition of pure public good is taken largely from Robert Bennett, *The Geography of Public Finance* (London: Methuen, 1980), pp. 11–13, but draws

also upon Paul Samuelson, 'The Pure Theory of Public Expenditure,' *Review of Economics and Statistics*, 36 (1954), pp. 387–9.

82 Mancur Olson, *The Logic of Collective Action* (Cambridge, Mass.: Harvard University Press, 1965).

83 Kevin Cox, *Conflict, Power and Politics in the City: A Geographic View* (New York: McGraw-Hill, 1973), p. 3; and Gordon Tullock, *Toward a Mathematics of Politics* (Ann Arbor: University of Michigan Press, 1967), pp. 82–3.

84 Kevin Cox, *op. cit.*, n. 83, pp. 10–11.

85 Robert Bennett, *op. cit.*, n. 81, pp. 64–5. What is more, neo-Keynesians have warned that territory can often be an end in itself. Aid is given to places which become surrogates for people. See M. Edel, ' "People" Versus "Places" in Urban Impact Analysis,' in N. Gluckman (ed.), *The Urban Impact of Federal Policies* (Baltimore: Johns Hopkins University Press, 1980), pp. 175–91.

86 Gordon Clark and Michael Dear, *State Apparatus: Structures and Language of Legitimacy* (Boston: Allen and Unwin, 1984), especially chapter 2.

87 This is a generalization of course because there are other names for these levels as well as other levels altogether.

88 Precisely what the role of the state is in capitalism has not been worked out. See Gordon Clark and Michael Dear, *op. cit.*, n. 86, especially chapter 2 for a review of the topic.

89 Gordon Clark, 'Towards a Critique of the Tiebout Hypothesis,' in A. Burnett and P. Taylor (eds.), *Political Studies from Spatial Perspectives* (New York: John Wiley, 1981), pp. 111–30; C. Tiebout, 'A Pure Theory of Local Expenditures,' *Journal of Political Economy*, no. 64 (1956), pp. 416–24. See also Ivan Szelenyi, 'The Relative Autonomy of the State or State Mode of Production,' in Michael Dear and Allen Scott (eds.), *Urbanization and Urban Planning in Capitalist Society* (New York: Methuen, 1981), pp. 565–91.

90 Richard Peet, 'Inequality and Poverty: A Marxist-Geographic Inquiry,' *Annals of the Association of American Geographers*, 65 (1975), pp. 564–71.

91 J. O'Connor, *The Fiscal Crisis of the State* (New York: St Martin's Press, 1973).

92 Kevin Cox, *Urbanization and Conflict in Market Societies* (Chicago: Maroufa Press, 1978), p. 84; and Ron Johnston, 'The Local State and the Judiciary,' in Robin Flowerdew (ed.), *Institutions and Geographic Patterns* (New York: St Martin's Press, 1982), especially p. 263.

93 Gordon Clark, 'The Law, the State, and the Spatial Integration of the United States,' *Environment and Planning A*, 13 (1981), pp. 1197–1232. Ref. on p. 1198.

94 *Ibid.*

95 *Ibid.*, p. 1212.

96 *Ibid.*, p. 1215.

97 A major use of territory in contemporary political organizations which is not addressed by any particular political theory but which follows from the theory of territoriality is the division of decision making into long and short range planning and responsibility and its association with levels of political territories. Many decisions in government use territory to divide the process into stages and levels. A plan can be formulated, approved, and funded at the national level, and then implemented and further funded at the state or local level. This is the case, for example, with nuclear power. The research, development, and national priorities

were established before the geographic details at the local level were worked out. By the time implemention was addressed there were few options left for the local units. The effect of such a process is to make the geographic impact precise at one level – the national – and to leave it imprecise or even undefined at another – the local – until much later when the process must be located elsewhere. To say the least, this procedure allows decisions to be made which might be strongly resisted at the local level.

6. The work place

1 Lyn Lofland, *A World of Strangers* (New York: Basic Books, 1973), pp. 30–2.
2 For the relationships between personal identity and geographical segmentation see Yi-Fu Tuan, *Segmented Worlds and Self* (Minneapolis: University of Minnesota Press, 1982).
3 The view is presented in most economic textbooks. For more technical discussions see A. Francis, J. Turk, and P. Willman, *Power, Efficiency and Institutions: A Critical Appraisal of the Markets and Hierarchies Paradigm* (London: Heinemann, 1983); and O. Williamson, *Markets and Hierarchies* (New York: Free Press, 1975). For an emphasis on the role of technology and the rise of the factory see D. Landes, *The Unbound Prometheus* (London: Cambridge University Press, 1969). The justification of management in particular is taken up by J. Litterer, 'Systematic Management: The Search for Order and Integration,' *Business History Review*, 35 (1961), pp. 461–76; J. Litterer, 'Systematic Management: Design for Organizational Recoupling in American Manufacturing Firms,' *Business History Review*, 37 (1963), pp. 369–91; and A. Chandler and H. Daems, *Managerial Hierarchies: Comparative Perspectives on the Rise of Modern Industrial Enterprise* (Cambridge: Harvard University Press, 1980).
4 Adam Smith, *An Inquiry into the Nature and Causes of the Wealth of Nations* (Chicago: Encyclopaedia Britannica, Great Books, 1952), pp. 3–4.
5 *Ibid.*, p. 4.
6 *Ibid.*, p. 4.
7 Frederick Engels, 'On Authority,' in L. Feuer (ed.), *Marx and Engels, Basic Writings in Politics and Philosophy* (New York: Doubleday and Co., 1959), p. 483, cited in S. Marglin, 'What Do Bosses Do? The Origins and Functions of Hierarchy in Capitalist Production,' *Review of Radical Political Economics*, 6 (1974–5), p. 33.
8 Michel Foucault, *Madness and Civilization* (New York: Random House, 1965), p. 46.
9 A typology of deviance based on work can be seen as early as 1601 in the Elizabethan Poor Law which distinguishes among the able bodied 'unemployed,' children, and the aged and impotent. In feudalism poverty was not a crime, see B. Tierney and P. Linehan (eds.), *Authority and Power* (Cambridge: Cambridge University Press, 1980); and Sidney and Beatrice Webb, *English Poor Law* (London: Longman, 1910).
10 Work would even cure the ill, Michael Ignatieff, *A Just Measure of Pain* (London: Macmillan, 1978).
11 Michel Foucault, *op. cit.*, n. 8, p. 51.

12 For reference to hospitals and healing in terms of efficiency see *ibid*.
13 Michael Ignatieff, *op. cit.*, n. 10, p. 61.
14 Cited in E. P. Thompson, 'Time, Work, Discipline, and Industrial Capitalism,' in M. Flinn and T. Smout (eds.), *Essays in Social History* (Oxford: Clarendon Press, 1974), p. 59. See also John Moffett, 'Bureaucratization and Social Control' (Columbus, Ohio: Ohio State University, Ph.D. thesis, 1971), p. 180. Early mass education was run like a factory. For factory image see Andrew Bell, *The Madras School*, 1808.
15 J. S. Hurt, *Elementary Schooling and the Working Classes, 1860–1918* (London: Routledge and Kegan Paul, 1978), p. 31, citing a Newcastle report of 1861. Nigel Thrift refers to this in chapter 3, p. 23, of 'Spontaneity or Order: A Social History of Time Consciousness,' vol. 1, *1300–1800* (unpublished manuscript).
16 Andrew Skull, 'A Convenient Place to Get Rid of Inconvenient People: The Victorian Lunatic Asylum,' in Anthony Kind (ed.), *Buildings and Society* (London: Routledge and Kegan Paul, 1980), p. 41; and Andrew Skull, *Museums of Madness* (London: Penguin, 1979).
17 Michel Foucault, *Discipline and Punish: The Birth of the Prison* (New York: Pantheon, 1977), p. 141.
18 *Ibid.*, pp. 141–2.
19 *Ibid.*, p. 143.
20 *Ibid.*, pp. 144–5.
21 *Ibid.*, p. 147.
22 Jeremy Bentham, *Panopticon* . . ., in John Bowring (ed.), *The Works of Jeremy Bentham* (Edinburgh: William Tait, 1843), preface. For a companion piece see Bentham's 'Outline of a Work, Entitled Pauper Management Improved,' in *ibid.*, vol. 8, pp. 369–439.
23 Jeremy Bentham, *Panopticon* . . ., in *ibid.*, preface.
24 *Ibid.*, p. 40.
25 *Ibid.*, p. 44.
26 *Ibid.*, p. 45.
27 *Ibid.*, p. 45.
28 *Ibid.*, pp. 40–1.
29 *Ibid.*, p. 44.
30 *Ibid.*, p. 45.
31 *Ibid.*, p. 45.
32 *Ibid.*, p. 42.
33 *Ibid.*, p. 60.
34 *Ibid.*, p. 62.
35 *Ibid.*, p. 63.
36 *Ibid.*, p. 61.
37 Robin Evans, *The Fabrication of Virtue: English Prison Architecture, 1750–1840* (Cambridge: Cambridge University Press, 1982), pp. 228–30. For other Panopticon designs see Negley Teeters and John Shearer, *The Prison at Philadelphia* (New York: Temple University Press, by Columbia University Press, 1957).
38 John Thompson and Grace Goldin, *The Hospital: A Social and Architectural History* (New Haven: Yale University Press, 1975).

39 Michel Foucault, *op. cit.*, n. 17, p. 145.
40 Sigfried Giedion, *Space, Time, and Architecture* (Cambridge, Mass., Harvard University Press, 1967), p. 184.
41 H. Roland (pseud. H. A. Arnold), 'Effective Systems of Finding and Keeping Shop Costs,' Part II, in 'The Collective Job-Ticket Adapted to a Drop-Forging Works,' *Engineering Magazine*, 15 (1898), p. 244. I am indebted to Richard Mahon for references in the *Engineering Magazine* and in *American Machinist*.
42 *Ibid.*, p. 244.
43 H. A. Arnold, 'Effective Systems for Finding and Keeping Shop Costs,' Part IV, in 'The Hyatt Roller Bearing Company's Practice,' *Engineering Magazine*, 15 (1898), pp. 749–58, 470–2.
44 'A Manufacturing Machine Shop,' *American Machinist* (July 16, 1896), p. 2.
45 *Ibid.*
46 *Ibid.*, p. 377.
47 *Ibid.*, p. 390.
48 Oswald W. Grube, *Industrial Buildings and Factories* (New York: Praeger Publishers, 1971), p. 14.
49 *Ibid.*, p. 9.
50 Robin Evans, 'Figures, Doors, and Passages,' *Architectural Design*, 48 (1978), pp. 267–78. See Robin Evans, 'Rookeries and Model Dwellings,' *Architectural Association Quarterly*, 10 (1978), pp. 24–35 for a discussion of internal partitioning of houses for the poor as an instrument of reform. For the elaborate partitioning of the upper class Victorian house, see Mark Girouard, *The Victorian Country House* (Oxford: Clarendon Press, 1971).
51 Giedion, *op. cit.*, n. 40, p. 363.
52 *Ibid.*, p. 378.
53 *Ibid.*, pp. 405 and 582.
54 Kitchens were often presented as little factories in which food was produced.
55 A classic review of the domestic system is found in Paul Mantoux, *The Industrial Revolution in the Eighteenth Century*, revised edn, and trans. by Marjorie Vernon (New York: Harcourt Brace Jovanovich, 1962). The system has also been called the putting out system when the cottager is linked to a merchant supplier who pays the cottager wages for work.
56 The freedom of the domestic worker should not be exaggerated. It was relative to the factory. The cottager may have had to work long hours, under hardships, and for little pay, but he could still pace his own work within the week and have a say in how his tasks would be organized.
57 Paul Mantoux, *op. cit.*, n. 55, pp. 60–1. See also Derek Gregory, *Regional Transformation and Industrial Revolution: A Geography of the Yorkshire Woollen Industry* (London: Macmillan, 1982), *passim*.
58 Derek Gregory, *op. cit.*, n. 57.
59 For Leicestershire see W. Hoskins and R. McKinley, *A History of the County of Leicester* (London: Oxford University Press, 1955), vol. 3, pp. 1–57, wherein is mentioned the role of baggers. For more detail about the conditions of framework knitting see D. M. Smith, 'The Location of the British Hosiery Industry Since the Middle of the 19th Century,' in R. Osborne and F. Barnes (eds.), *Geographical Essays in Honour of K. C. Edwards* (Nottingham: Univer-

sity of Nottingham, Department of Geography, 1970), pp. 71–9; and A. Pickering, *The Cradle and Home of the Hosiery Trade* (Hinckley, 1940).

60 Paul Mantoux, *op. cit.*, n. 55, p. 34, says that 'the class of manufactories which Jack of Newbury represents developed rapidly during the first half of the sixteenth century.' Yet there is little evidence to point to how large a percentage of the weaving actually took place in such establishments. It seems that overall the process was a cottage industry until the introduction of steam. See Albert Usher, *An Introduction to the Industrial History of England* (Boston: Houghton Mifflin, 1920); and D. Bythell, *The Handloom Weavers* (Cambridge: Cambridge University Press, 1969). See also S. Marglin, *op. cit.*, n. 7, p. 46.

61 Paul Mantoux, *op. cit.*, n. 55, pp. 31–3. See also H. Levy, *Monopoly and Competition* (London: Macmillan, 1911); W. Ashley, *Introduction to English Economic History and Theory* (New York: A. M. Kelley, 1966), vol. 2; and G. Unwin, *Industrial Organization in the Sixteenth and Seventeenth Centuries* (Oxford: Clarendon Press, 1904).

62 The most direct sources for working conditions in the textile industry are found in the Parliamentary Papers on the weaving industry.

63 E. Nelson, 'The Putting Out System in the English Framework Knitting Industry,' *Journal of Economics and Business History*, 2 (1930), pp. 467–94.

64 Economics of the factory are clearly outlined in R. Millward, 'The Emergence of Wage Labour in Early Modern England: Exploration of An Analytical Framework,' manuscript (Manchester: University of Salford, Department of Economics, 1980).

65 Sidney Pollard, *The Genesis of Modern Management* (Cambridge, Mass.: Harvard University Press, 1965), p. 252. Andrew Ure, *The Philosophy of Manufacturers* (London: Charles Knight, 1835), describes the major difficulties that Arkwright found as not lying 'so much in the invention of a proper self-acting mechanism for drawing out and twisting cotton into a continuous thread, as in . . . [devising and administering] a successful code of factory discipline, suited to the necessities of factory diligence . . . It required . . . a man of a Napolean nerve and ambition, to subdue the refractory tempers of workpeople, accustomed to irregular paroxysms of diligence' (pp. 15–16).

66 A. Friedman, *Industry and Labour* (London: Macmillan, 1977), p. 152; Sidney Pollard, *op. cit.*, n. 65, agrees that the capitalists' apprehensions over centralizing workshops were outweighed by their advantages in efficiency of control. Specific reasons for placing framework knitters together are given in Parliamentary Papers and Reports on the Conditions of the Framework Knitters; see especially App. I and II for Hinckley.

67 See Derek Gregory, *op. cit.*, n. 57, for a discussion of the various interests in and forms of resistance to the factory system in the Yorkshire woolen industry.

68 Paul Mantoux, *op. cit.*, n. 57, p. 35.

69 Derek Gregory, *op. cit.*, n. 57, p. 130.

70 G. Unwin, *Samuel Oldknow and the Arkwrights* (Manchester: Manchester University Press, 1924), pp. 34–5, citing acts of 1777 of the West Riding, and later acts of 1784, 1785, and 1790 for other areas. 'There was not much left of the independence of the small master except the choice of his hours of labour . . .' (p. 35).

71 John Prest, *The Industrial Revolution in Coventry* (London: Oxford University Press, 1960), p. 45.
72 *Ibid.*, p. 96.
73 See Parliamentary Papers on the factory and Karl Marx.
74 Harry Braverman, *Labor and Monopoly Capital: The Degradation of Work in the Twentieth Century* (New York: Monthly Review Press, 1974), p. 66.
75 Sidney Pollard, *op. cit.*, n. 65, p. 163.
76 *Ibid.*, pp. 164–5. On p. 69, Pollard states: 'The thoughts of the early entrepreneurs, looking for docile labour of a new kind, turned easily to unfree labour, both here and on the Continent. Nor were the complementary efforts to turn poorhouses into workhouses–manufactories entirely unconnected, for the aims to see idle men punished and educated to work, and to lower the poor rates, were aims of the employing classes everywhere.'
77 For an overview of the scattered references in the debate about the emiseration of the working poor see Eric Hobsbawm, *Industry and Empire* (London: Weidenfeld and Nicolson, 1968).
78 N. Gras, *Industrial Evolution* (Cambridge: Harvard University Press, 1930), quoted in Harry Braverman, *op. cit.*, n. 74, p. 66.
79 Harry Braverman, *op. cit.*, n. 74, pp. 65–6.
80 S. Marglin, *op. cit.*, n. 7, p. 38.
81 *Ibid.*, p. 40.
82 *Spectator*, London, May 26, 1866, p. 569, quoted in *ibid.*, p. 40.
83 Harry Braverman, *op. cit.*, n. 74, p. 195.
84 *Ibid.*, p. 195, quoting Charles Babbage, *On the Economy of Machinery and Manufactures* (New York, 1963), p. 54.
85 Richard Edwards, *Contested Terrain* (New York: Basic Books, 1979), pp. 110–14.
86 Mathew L. Wald, 'Back Offices Disperse from Downtown: Technology Spurs Decentralization Across the Country,' *New York Times*, Section 12, May 13, 1984, pp. 1 and 58–61, discusses recent decentralizing effects of technology on office work.
87 One can add also that having workers work at home, but on instruments owned and monitored by others, can run the risk of turning the home into something like a factory. It can also lead to employers entering homes to inspect equipment. This could be reminiscent of the invasions of privacy that cottage workers endured.
88 'United States Army Infantry Reference Data I and II' (Fort Benning, Georgia: Department of the Army, U.S. Army Infantry School, ST7–157 FY72, 1972); and 'The Role of the U.S. Army Special Forces' (Washington, D.C.: Headquarters, Department of the Army, TC–31–20–1, 1976).
89 S. Sarkesian (ed.), *Combat Effectiveness* (Beverly Hills: Sage Publishing, 1980); and J. Willer, *Fire and Movement* (New York: Crowell and Co., 1967).
90 'United States Army Infantry Reference Data I and II' and 'The Role of the U.S. Army Special Forces,' *op. cit.*, n. 88.
91 Arthur Turner and Paul Lawrence, *Industrial Jobs and the Worker* (Cambridge, Mass.: Harvard University Press, 1965), as summarized by Robert Cooper,

'Task Characteristics and Intrinsic Motivation,' *Human Relations*, 26, no. 3 (1973), pp. 387–413.

92 Robert Cooper, *op. cit.*, n. 91, p. 390. For other studies of industrial organization in which several of our variables are addressed, see J. Kimberly, 'Organizational Size and the Structuralist Perspectives: A. Review, Critique and Proposal,' *Administrative Science Quarterly*, 21 (1976), pp. 571–97; Joan Woodward, *Industrial Organization: Theory and Practice* (London: Oxford University Press, 1965); and Joan Woodward (ed.), *Industrial Organization: Behaviour and Control* (London: Oxford University Press, 1970). See also the *Dictionary of Occupational Titles (DOT)*, published by the Department of Labor and the scores for each occupation on forty-four characteristics. These are available by writing to National Technical Information Services, 5285 Port Royal Road, Springfield, Virginia 22161, quoting document no. PB 298 315/AS. See also their no. ADA 057268 which is the G. Oldham and J. Hackman 'Norms for the Job Diagnostic Survey.'

93 Several of the studies cited above suggest that there is a point where too much complexity and discretion in tasks can create worker dissatisfaction. See also n. 100.

94 *Work in America: Report of a Special Task Force to the Secretary of Health, Education, and Welfare* (Cambridge, Mass.: M.I.T. Press, 1973).

95 For studies and proposals on work redesign see J. Hackman and G. Oldham, *Work Redesign* (Reading, Mass.: Addison-Wesley, 1980); P. Dickson, *The Future of the Workplace* (New York: Webright and Talley, 1975); F. Emery and E. Thorsrud, *Form and Content in Industrial Democracy* (London: Tavistock, 1969); D. Pugh and R. Payne, *Organizational Behavior in its Context: The Aston Programme III* (Westmead: Saxon House, 1977).

96 John Logue, 'SAAB/Trollhatan: Reforming Work Life on the Shop Floor,' *Working Life in Sweden*, no. 23 (June 1981), p. 3 (obtainable from Swedish Information Service, Swedish Consulate General, 825 Third Ave., N.Y. 10022, U.S.A.).

97 *Ibid.*, p. 3.

98 *Ibid.*, p. 3.

99 *Ibid.*, p. 4.

100 Berth Honsson, 'The Volvo Experiences of New Job Design and New Production Technology,' *Working Life in Sweden*, no. 19 (Sept. 1980), p. 2. Yet the innovations may have increased stress in some cases. See Bertil Gardfell, 'Stress and Its Implications in Sweden,' *Working Life in Sweden*, no. 20 (Oct. 1980), p. 3.

101 D. Kingdon, *Matrix Organization* (London: Tavistock, 1973).

102 Einar Thorsrud, 'Democratization of Work as a Process of Change: Towards Non-Bureaucratic Types of Organizations,' in Geert Hofstede and M. Kassem (eds.), *European Contributions to Organization Theory* (Amsterdam: Van Corcum, 1976), p. 263.

103 For architecture and function in schools see L. Krasner, 'The Classroom as a Planned Environment,' *Educational Researcher*, 5 (1976), pp. 9–14.

104 Greg Oldham and Daniel Brass, 'Employee Reactions to an Open-Plan Office: A Naturally Occurring Quasi-Experiment,' *Administrative Science Quarterly*,

24 (1979), pp. 267–84. For other studies of the effects of office design on behavior see M. J. Brookes and A. Kaplan, 'The Office Environment: Space Planning and Effective Behavior,' *Human Factors*, 14 (1972), pp. 373–91; D. Canter, 'Reactions to Open Plan Offices,' *Built Environment*, 1 (1972), pp. 465–7; A. Hundert and N. Greenfield, 'Physical Space and Organizational Behavior: A Study of Office Landscape,' *Proceedings of the 77th Annual Convention of the American Psychological Association*, 1 (1969), pp. 601–2; and Peter Manning (ed.), *Office Design: A Study of Environment* (Liverpool: University of Liverpool, Pilkington Research Unit, Dept of Building Science, 1965).

105 *Ibid.*, p. 273.

106 *Ibid.*, p. 280.

107 'Wearing a Jail Cell Around Your Ankle,' *Newsweek*, March 21, 1983, p. 53.

Author Index

Bracketed number refers to page in Notes where citation is given.

Subject Index